COMPUTER SOLUTION
OF ORDINARY
DIFFERENTIAL EQUATIONS

COMPUTER SOLUTION
OF ORDINARY
DIFFERENTIAL EQUATIONS
The Initial Value Problem

L. F. SHAMPINE / M. K. GORDON

SANDIA LABORATORIES AND UNIVERSITY OF NEW MEXICO

W. H. FREEMAN AND COMPANY
San Francisco

Library of Congress Cataloging in Publication Data

Shampine, Lawrence F
    Computer solution of ordinary differential equations

    Bibliography: p.
    Includes index.
    1. Electronic data processing—Differential equations—Numerical solutions. 2. Electronic data processing—Initial value problems—Numerical solutions
I. Gordon, Marilyn Kay, joint author. II. Title.
QA372.S416   515′.352′02854   74-23246
ISBN 0-7167-0461-7

Printed in the United States of America

1 2 3 4 5 6 7 8 9

CONTENTS

PREFACE

This book shows how to solve differential equations on a digital computer. It has become traditional for texts on the subject to survey the basic theory of the various methods. In contrast, this text treats as completely as possible one method, generally conceded to be the most effective when properly used. Practical aspects of the algorithms and their implementation are just as important to efficiency and reliability as the choice of the fundamental method itself. Such questions have been largely ignored outside the research literature. As a consequence, much of the theory that we present has been taken directly from the literature or developed for this text. By concentrating on a single method we are able to give a treatment which is both simple and far more complete than has been given before. The codes developed are among the most effective and the most carefully justified and documented available at this time.

A great many methods have been proposed for the solution of differential equations, but relatively few have been put into practice. There are very few codes readily available that have been carefully written for general use and represent the state of the art at the time that they were written. The codes DVDQ of F. T. Krogh [20] and DIFSUB of C. W. Gear [3] are responsible for the recognition of the effectiveness of a variable order, variable step formulation of the classic Adams methods. Solving differential equations in this way is the subject of this text, and the authors are glad to acknowledge their debt to these codes and to the researches of their writers. When the equation is expensive to evaluate, this approach appears to be more efficient by far than any other general purpose method. In addition, it is easy to detect and cope with a variety of abnormal problems, so that carefully written codes of this kind are relatively safe. The most prominent competing methods are the Runge-Kutta methods and the extrapolation methods. Of the three families of methods, the fixed order Runge-Kutta is the simplest, in several respects the best understood, and the least efficient. There are a number of features of the Adams methods which are not available with this family, but if one ignores them, and decides against providing others, a very simple, compact code can be written. The development of this topic in the elementary text [25] is a valuable preparation for the more involved study contained herein. The variable order extrapolation methods have an efficiency approaching that of the Adams methods but are the least understood at this time, especially as regards practice. Again, realizing that various features of the Adams methods will not be present, one can write codes of intermediate complexity. Old or entirely new methods may well be developed that will supplant the Adams methods. Since this book is not a survey, it will not assist researchers seeking to do this, but it will provide a standard for judging their success.

The book may be read and used for several purposes. Readers interested only in using the codes to solve problems need little more in the way of background than an ability to read FORTRAN programs easily. Chapter 1 should be read for general information, as should the material on assessing global errors in Chapter 7. That part of Chapter 10 explaining how to use the codes, and all of Chapter 11, which illustrates how the codes perform on both normal and abnormal problems, should also be carefully read. Chapter 12, with its examples and techniques for solving differential equations, will also be useful.

As a text, the book provides a self-contained treatment of an important problem. It also shows how the numerical analyst must blend abstract theory, analysis of special situations and model problems, heuristics, and careful experimentation to produce effective codes. The demands on the reader are rather modest. A command of the calculus and reasonable proficiency with FORTRAN will suffice. Some familiarity with differential equations and elementary matrix theory is desirable, though not essential. Naturally, a thorough understanding of the whole book requires a certain amount of sophistication and experience in both mathematics and computing. One intended use as a textbook is for a topics course in numerical analysis. For this purpose the book should be read straight through. Another use is as a supplementary text to cover the solution of differential equations by digital computer for a survey course on numerical analysis. For this purpose certain topics should be skimmed and others skipped. Chapters 1, 2, and 3 should be read carefully for a basic understanding of differential equations and their solution. The first part of Chapter 4 should be read in order to understand what is meant by convergence when the step size and order are fixed, and the last part just skimmed to see that varying the step size and order leads to much the same results. Chapter 5 may be skipped if the reader appreciates that it just extends the classical backward difference implementation, discussed in Chapter 3, to variable step size. However, the computer scientist might be interested in this chapter, because it is largely devoted to how one efficiently organizes and programs the algorithm. The first part of Chapter 6 should be read to see, more precisely than is shown in Chapter 3, what errors are being measured in the codes, and to appreciate what their control implies. Also, the important matter of local extrapolation is explained. Beyond this point the chapter can be skipped. Chapter 7 is particularly illustrative of the art of numerical analysis, as it shows how one uses some theory, model problems, and experimentation to arrive at effective algorithms.

When used as a text, the reading outlined above should be paralleled with practical experience in using the codes and solving problems. As soon as possible, one should read that part of Chapter 10 describing how to use the codes, and follow this by examination of the results reported in Chapter 11, and experiments of one's own. There are some exercises scattered throughout the text that serve to test the reader's understanding and sometimes to introduce supplementary material. They, and their solutions, should at least be examined. The principal source of exercises is Chapter 12. This chapter attempts to pass on, by example and exercise, some of the art of solving differential equations. Readers should test their understanding of the theory and practice of solving differential equations by coordinating this chapter with their own reading and experimentation plans.

The authors are very grateful to S. M. Davenport for her many contributions to this book, particularly in the development and testing of the codes, and to W. Gautschi, W. B. Gragg, and R. J. Thompson for their reviews of early manuscripts which helped shape this final version. The authors acknowledge their intellectual debt to the published work and private communications of F. T. Krogh, C. W. Gear, and T. E. Hull. These eminent workers in the field have greatly influenced our thinking.

The codes have been field tested at a number of laboratories, and have been considerably improved as a result. Our colleagues at Sandia

Laboratory Albuquerque, Sandia Laboratory Livermore, Lawrence Livermore Laboratory, Los Alamos Scientific Laboratory, and Argonne National Laboratory have been particularly helpful.

We are grateful to Sandia Laboratories and the University of New Mexico for their support of this project in many ways and, in particular, for the services of two excellent technical typists, B. P. Vigil and M. H. Richardson, who were very important to the production of this book.

July 1974   *L. F. Shampine*
                                                              M. K. Gordon

COMPUTER SOLUTION
OF ORDINARY
DIFFERENTIAL EQUATIONS

1

FUNDAMENTALS
OF THE THEORY

A differential equation is an equation involving a function and its derivatives. In the calculus one obtains the solutions of a number of differential equations, although they are not always described as such. For example, if $f(x)$ is a function continuous on an interval $[a, b]$, the equation

$$\frac{dy}{dx} = f(x) \tag{1}$$

has a solution given by the fundamental theorem of calculus:

$$y(x) = B + \int_a^x f(t)\, dt. \tag{2}$$

Since this represents a solution for any value of the constant B the equation alone is not enough to specify a particular solution. One way to select a particular solution from the family of solutions (2) is to give its value at a, the initial point:

$$y(a) = A. \tag{3}$$

We call the problem posed by (1) and (3) an initial value problem and find from (2) that it has exactly one solution, namely

$$y(x) = A + \int_a^x f(t)\, dt.$$

There are other ways to specify a particular solution, but giving an initial value is by far the most common. In general we mean by an initial value problem for a first order ordinary differential equation a problem of the form

$$\frac{dy}{dx} = f(x, y), \tag{4}$$

$$y(a) = A. \tag{5}$$

We shall always presume that $f(x, y)$ is continuous in both variables. By a solution $y(x)$ of (4) and (5) we mean that on some interval $[a, b]$, $y(x)$ is continuous and has a continuous derivative, $y(x)$ satisfies (5), and

$$\frac{dy(x)}{dx} = f(x, y(x)) \qquad \text{for each } x \text{ in } [a, b].$$

A good bit of insight can be gleaned from equations of the special form

$$\frac{dy}{dx} = f(y) \tag{6}$$

which can be "solved" in a simple way. If $f(A) \neq 0$, say $f(A) > 0$, equation (6) says that $y(x)$ has to increase as long as $f(y)$ remains positive. But then the inverse function $x(y)$ exists and is relatively easy to obtain. Recall from the calculus that if dy/dx is continuous and non-zero on some interval in x, then the inverse function $x(y)$ exists on the corresponding interval in y and

$$\frac{dx(y)}{dy} = \frac{1}{dy(x)/dx}.$$

If $f(A) \neq 0$, we can solve the problem

$$\frac{dx}{dy} = \frac{1}{f(y)},$$

$$x(A) = a$$

to obtain its unique solution

$$x(y) = a + \int_A^y \frac{dt}{f(t)}.$$

As long as $f(y) \neq 0$ this defines $x(y)$ and its inverse function $y(x)$ which is the solution of (5) and (6). By way of a familiar example consider

$$\frac{dy}{dx} = y,$$

$$y(0) = A$$

for $A > 0$. The general solution just obtained yields in this case

$$x(y) = \int_A^y \frac{dt}{t} = \ln(y/A)$$

or

$$y(x) = Ae^x,$$

valid for all x. The same conclusion holds for $A < 0$ because then $y < 0$.
  The example

$$\frac{dy}{dx} = y^2,$$

$$y(0) = 1$$

shows why the definition of a solution is made with respect to an interval. For

$$x(y) = \int_1^y \frac{dt}{t^2} = 1 - \frac{1}{y}$$

hence

$$y(x) = \frac{1}{1-x}.$$

The solution does not exist on any interval containing $x = 1$ since it "blows up" there.

If $f(A) = 0$, then $y(x) \equiv A$ is always a solution of (5) and (6). It is not clear whether there are other solutions since the solution technique just developed does not apply. An example with several solutions is

$$\frac{dy}{dx} = y' = -\sqrt{|1 - y^2|}, \tag{7}$$

$$y(0) = 1. \tag{8}$$

In addition to the obvious solution $y(x) \equiv 1$ there is also a solution $\cos x$ on a suitable interval. For if $y(x) = \cos x$, then

$$y' = -\sin x$$

and from equation (7) and the identity $\sin^2 x + \cos^2 x = 1$,

$$y' = -\sqrt{|1 - \cos^2 x|} = -|\sin x|.$$

This says $\cos x$ satisfies the differential equation for $0 \le x \le \pi$ and, of course, $\cos 0 = 1$. Notice that for $\pi < x < 2\pi$, $\cos x$ does not satisfy the differential equation. However, the function

$$y(x) = \begin{cases} \cos x & 0 \le x \le \pi \\ -1 & x > \pi \end{cases}$$

is a solution on $[0, b]$ for any $b > 0$ (we write this as $[0, \infty)$). Obviously both $\cos x$ and -1 satisfy the differential equation, and checking at $x = \pi$ shows that $y(x)$ is continuous and has a continuous derivative at that point, hence everywhere. With this as a guide we discover that there exists an infinite number of solutions. For any number $\alpha \ge 0$ the function

$$y(x) = \begin{cases} 1 & 0 \le x \le \alpha \\ \cos(x - \alpha) & \alpha < x \le \alpha + \pi \\ -1 & \alpha + \pi < x \end{cases}$$

is a solution of the problem valid on $[0, \infty)$. (See Figure 1.)

EXERCISE 1

Verify that for any number $\alpha \le 0$,

$$y(x) = \begin{cases} 1 & \alpha \le x \le 0 \\ \cosh(\alpha - x) & x < \alpha \end{cases}$$

is a solution of (7) and (8) valid on $(-\infty, 0]$.

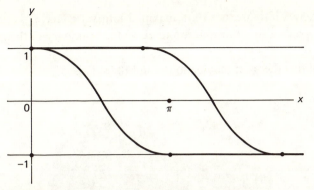

FIGURE 1

An initial value problem can have many solutions. The problem $y' = -\sqrt{1 - y^2}$, $y(0) = 1$ has infinitely many solutions; two are sketched in this figure.

EXERCISE 2

An equation of the form

$$y' = g(x)y + r(x)$$

is said to be linear. If $g(x)$ and $r(x)$ are continuous for $a \leq x \leq b$, verify that the solution of the initial value problem with $y(a) = A$ is

$$y(x) = u(x)[A + v(x)]$$

where

$$u(x) = \exp\left[\int_a^x g(t)\,dt\right],$$

$$v(x) = \int_a^x r(t)/u(t)\,dt.$$

Few initial value problems, even those of the special form (6), have analytical solutions in terms of common functions. For purposes of illustration we have chosen examples for which the analytical solution is given by a well-known function but the reader should not be misled. In general, even the existence of a solution must be deduced from the differential equation itself, not by finding an analytical expression, and the solution must be evaluated numerically. The following theorem is a fundamental result establishing the existence of a solution for initial value problems.

THEOREM 1 If $f(x, y)$ is continuous on an open plane region R containing the point (a, A), there is at least one solution of (4) and (5) in the region. Each solution exists on a maximal interval $\alpha < x < \beta$ containing a, and as $x \to \alpha$ or $x \to \beta$ the solution tends to the boundary of the region.

A proof of this theorem may be found in $[\mathbf{10}, \text{p. } 17]$.

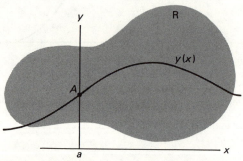

FIGURE 2

If $f(x, y)$ is continuous in both variables in a region R, then for any point (a, A) in R, the initial value problem $y' = f(x, y)$, $y(a) = A$ has a solution $y(x)$ that extends to the boundary of R (Theorem 1).

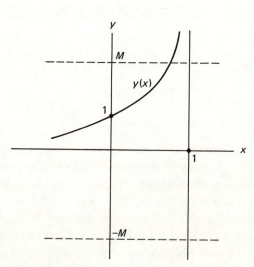

FIGURE 3

An initial value problem may not have solutions that extend to a given point. The problem $y' = y^2$, $y(0) = 1$ has only the solution $y(x) = 1/(1 - x)$, which becomes infinite as x tends to 1.

Notice how the theorem applies to the example with solution $1/(1-x)$. If we say the region R is $\{(x, y) | -2 < x < 2, -M < y < M\}$ for some $M > 1$, the solution reaches the boundary of the region when $x = 1 - 1/M$ and $x = -2$, just as the theorem states. We have seen by example that the theorem in no way guarantees a unique solution. Still, few problems of physical interest exhibit these kinds of behavior. A problem is said to be well posed if small changes in the initial value and the differential equation cause only small changes in the solution. When it describes a physical situation, a problem can only be an approximation, an idealization, because measurements can be made only with a limited accuracy. To be useful, an equation describing a physical problem must be well posed, so that small errors of measurement do not affect the solution too seriously. In fact, any problems which we attempt to solve numerically must be well posed since errors in their solution are inevitable. Fortunately a very simple condition guarantees that a problem is well posed. If $\partial f/\partial y$ exists and is bounded in the region of interest, then the problem is well posed. A condition which is less restrictive, but still convenient, will be used in all of our analytical studies. A function $f(x, y)$ is said to satisfy a Lipschitz condition (in y) with constant L in a region R if

$$|f(x, u) - f(x, v)| \le L|u - v| \qquad \text{for all } (x, u), (x, v) \text{ in } R.$$

For example, we say the function $f(x, y)$ is linear if

$$f(x, y) = g(x)y + r(x).$$

If $g(x)$ and $r(x)$ are continuous on the finite interval $[a, b]$, then $f(x, y)$ satisfies a Lipschitz condition in the region

$$R = \{(x, y) | a \le x \le b, -\infty < y < \infty\}.$$

This is true since

$$|f(x, u) - f(x, v)| = |g(x)| |u - v| \le L|u - v|$$

if L is any constant such that

$$\max_{a \le x \le b} |g(x)| \le L.$$

The continuity of $g(x)$ implies it is bounded so that an L exists. This example is a special case of the observation that if $f_y(x, y)$ is continuous and bounded in R, then $f(x, y)$ satisfies a Lipschitz condition. This is seen from

$$|f(x, u) - f(x, v)| = |f_y(x, \xi)| |u - v| \le L|u - v|,$$

if L is any constant such that

$$\max_{R} |f_y(x, y)| \le L.$$

EXERCISE 3

Prove that if f and f_y are continuous in a region R but f_y is not bounded there, then f cannot satisfy a Lipschitz condition in R. Hint: Suppose f does satisfy a Lipschitz condition with constant L. Using the limit definition of the derivative, prove that $|f_y| \le L$.

EXERCISE 4

Which of the following equations satisfy a Lipschitz condition in the region $|x| < \infty, |y| < \infty$? Find a suitable Lipschitz constant when this is possible.
 a) $y' = y/(1 + x^2)$,
 b) $y' = xy$,
 c) $y' = y \sin x$,
 d) $y' = \sqrt{|y|}$,
 e) $y' = \sqrt{|x|}$.

We shall now prove that Lipschitzian problems are well posed. The readers who have difficulties with the mathematics can skip the proofs of the theorems in this chapter; they will not suffer for it if they understand what the theorems say.

THEOREM 2 Suppose that $f(x, y)$ and $g(x, y)$ are continuous on an open region R and that $f(x, y)$ satisfies a Lipschitz condition with constant L there. Suppose further that

$$|f(x, y) - g(x, y)| \le \varepsilon \qquad \text{for all } (x, y) \text{ in } R.$$

If, for $a \le x \le b$,

$$u'(x) = f(x, u(x)),$$
$$v'(x) = g(x, v(x)),$$

and $(x, u(x)), (x, v(x))$ lie in R, then

$$|u(x) - v(x)| \le \{|u(a) - v(a)| + (b - a)\varepsilon\} e^{L(x - a)}.$$

Proof. First we integrate the equations to get

$$u(x) = u(a) + \int_a^x u'(t)\,dt = u(a) + \int_a^x f(t, u(t))\,dt$$

and a similar expression for $v(x)$. Subtracting $v(x)$ from $u(x)$, with some additional manipulation, gives

$$u(x) - v(x) = u(a) - v(a) + \int_a^x \{f(t, v(t)) - g(t, v(t))\}\,dt$$

$$+ \int_a^x \{f(t, u(t)) - f(t, v(t))\}\,dt.$$

Using the relation between f and g and the Lipschitz condition we see that

$$|u(x) - v(x)| \le |u(a) - v(a)| + \int_a^x \varepsilon\,dt + \int_a^x L|u(t) - v(t)|\,dt$$

$$\le |u(a) - v(a)| + (b - a)\varepsilon + \int_a^x L|u(t) - v(t)|\,dt \qquad \text{for } a \le x \le b.$$

Let us use the notation

$$\Delta(x) = |u(x) - v(x)|,$$

$$R(x) = |u(a) - v(a)| + (b - a)\varepsilon + \int_a^x L|u(t) - v(t)|\,dt.$$

In this notation the inequality just proved is

$$\Delta(x) \le R(x).$$

Since $R'(x) = L\Delta(x)$ the inequality implies

$$0 \ge R'(x) - LR(x)$$

which, because of the positivity of the exponential, implies

$$0 \ge e^{-L(x-a)}[R'(x) - LR(x)] = \frac{d}{dx}[R(x)\,e^{-L(x-a)}].$$

This means that the function $R(x)\,e^{-L(x-a)}$ does not increase, so

$$R(x)\,e^{-L(x-a)} \le R(a)\,e^{-L(a-a)} = R(a).$$

This leads to

$$\Delta(x) \le R(x) \le R(a)\,e^{L(x-a)}$$

which is just a different notation for the desired result

$$|u(x) - v(x)| \le \{|u(a) - v(a)| + (b - a)\varepsilon\} \, e^{L(x-a)}.$$

COROLLARY 1   Suppose that $f(x, y)$ is continuous on an open region R and satisfies a Lipschitz condition there. For any point (a, A) in the region R, the initial value problem $y' = f(x, y)$, $y(a) = A$ has at most one solution.

Proof.   If there were two solutions $u(x)$ and $v(x)$, we would have $u(a) = v(a) = A$ and $\varepsilon = 0$ in the theorem hence $u(x) - v(x) \equiv 0$.

Theorem 2 says that Lipschitzian problems are well posed, since if the initial condition is changed a little and the equation is changed a little, the solution is changed only a little. It also implies that a solution is a continuous function of the initial value, since if $u(x)$ and $v(x)$ both satisfy the equation $y' = f(x, y)$, then

$$|u(x) - v(x)| \le |u(a) - v(a)| \, e^{L(x-a)}.$$

In fact, if f_y exists and is continuous, the solution is a differentiable function of the initial value. If $y(x, A)$ is the solution with the initial value A at $x = a$ and if we let

$$u(x) = \frac{\partial y(x, A)}{\partial A},$$

then $u(x)$ is the solution of the linear differential equation

$$u' = f_y(x, y(x, A))u,$$

$$u(a) = 1,$$

which is called the equation of first variation. This is what one finds by formally differentiating the differential equation satisfied by $y(x, A)$ and interchanging the order of differentiations. It is not difficult to prove that this formal derivation is justified but it is rather out of our way so we omit it; a proof can be found in [24, pp. 136–140].

With a little strengthening of the hypotheses in the preceding theorem we can prove that the solutions will not "blow up."

THEOREM 3   Suppose $f(x, y)$ is continuous and satisfies a Lipschitz condition on an open region containing $\{(x, y) \,|\, a \le x \le b, \, -\infty < y < \infty\}$. Then for any A the problem $y' = f(x, y)$, $y(a) = A$ has a unique solution existing on all of the finite interval $[a, b]$.

Proof. Theorem 1 states that there is a solution which extends to the boundary of the region and the corollary to Theorem 2 states that there is only one solution. We only need to prove that the solution extends as far as $x = b$ and does not become infinite before b is reached. The basic argument is to suppose that the solution exists as far as $x = \alpha$. We then consider a rectangle $R = \{(x, y) \mid \alpha \leq x \leq \alpha + \Delta, -M < y < M\}$ for suitable positive Δ and (finite) M. The solution must extend to the boundary of this rectangle since R will be chosen to lie in the region; we shall prove that it does not go out the top or the bottom of the rectangle and hence must reach the side at $x = \alpha + \Delta$. Repetition of the basic argument shows that the solution extends as far as $x = b$.

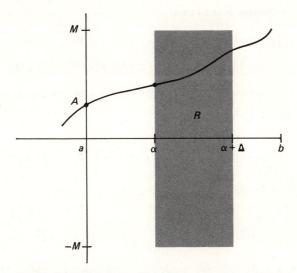

FIGURE 4

The basic construction for proving Theorem 3 assumes that the solution extends to $x = \alpha$. A rectangle R is constructed, and by Theorem 1 the solution extends to the boundary of R. It is shown that the solution cannot approach the top or bottom of R, hence it must reach the side at $x = \alpha + \Delta$. Since Δ is bounded away from zero, repetition of the argument shows that the solution extends to $x = b$.

For the basic argument we shall take $\Delta = \min(0.1/L, b - \alpha)$. To define M we first define

$$C = \max_{a \leq x \leq b} |f(x, 0)|$$

(which exists because f is continuous on this finite interval). Let γ be any number such that

$$0.1\, e^{0.1} < \gamma < 1$$

(this is possible since $0.1\, e^{0.1} < 1$). Then let

$$M = \left(|y(\alpha)| + \frac{C}{L} 0.1\, e^{0.1}\right)\Big/(\gamma - 0.1\, e^{0.1}).$$

To prove that $y(x)$ goes out the side of the rectangle we bound how far it can deviate from the constant function $v(x) \equiv y(\alpha)$, using Theorem 2. Now $v(x)$ is a solution of $v' = g(x, v)$, $v(\alpha) = y(\alpha)$ if we define $g(x, v) \equiv 0$. For (x, w) in the rectangle R we have

$$|f(x, w) - g(x, w)| = |f(x, w) - f(x, 0) + f(x, 0)|$$
$$\leq |f(x, w) - f(x, 0)| + |f(x, 0)|$$
$$\leq L|w| + C \leq LM + C.$$

All the conditions of Theorem 2 are satisfied and the conclusion is that

$$|y(x) - v(x)| \leq \Delta(LM + C)\, e^{L\Delta} = \left(M + \frac{C}{L}\right) L\Delta\, e^{L\Delta}$$

hence

$$|y(x)| \leq |y(\alpha)| + \left(M + \frac{C}{L}\right) 0.1\, e^{0.1} = \gamma M.$$

For $\alpha \leq x \leq \alpha + \Delta$, $y(x)$ is bounded in magnitude by $\gamma M < M$, so it does not reach the top or bottom of R. Since it does reach the boundary, it must go out the side at $x = \alpha + \Delta$.

SYSTEMS AND HIGHER ORDER EQUATIONS

So far we have considered a single equation but usually we deal with several at the same time. For example, $y_1(x) = \sin x$ and $y_2(x) = \cos x$ are the solution of

$$y_1' = y_2, \qquad y_1(0) = 0,$$
$$y_2' = -y_1, \qquad y_2(0) = 1.$$

The general initial value problem is

$$y_1' = f_1(x, y_1, y_2, \ldots, y_n), \qquad y_1(a) = A_1,$$
$$y_2' = f_2(x, y_1, y_2, \ldots, y_n), \qquad y_2(a) = A_2,$$
$$\vdots \qquad\qquad\qquad \vdots$$
$$y_n' = f_n(x, y_1, y_2, \ldots, y_n), \qquad y_n(a) = A_n.$$

In vector notation with

$$\mathbf{y} = (y_1, y_2, \ldots, y_n)^T,$$
$$\mathbf{A} = (A_1, A_2, \ldots, A_n)^T,$$
$$\mathbf{f}(x, \mathbf{y}) = (f_1(x, \mathbf{y}), f_2(x, \mathbf{y}), \ldots, f_n(x, \mathbf{y}))^T$$

this is

$$\mathbf{y}' = \mathbf{f}(x, \mathbf{y}),$$
$$\mathbf{y}(a) = \mathbf{A}.$$

The formal resemblance to the case of one equation is not misleading. All the preceding theorems remain true and their proofs require only minor changes.

Frequently we must consider differential equations involving derivatives of orders greater than one, e.g., the function $y(x) = \sin x$ satisfies the problem

$$y'' = -y, \qquad y(0) = 0, \qquad y'(0) = 1$$

involving second order derivatives.

For typical numerical procedures it is necessary to transform such problems into systems of first order equations by introducing new unknowns. In this example let $y_1(x)$ be the desired function $y(x)$, and introduce the new unknown $y_2 = y_1'$. Then

$$y_1' = y_2, \qquad y_1(0) = 0,$$
$$y_2' = y_1'' = -y_1, \qquad y_2(0) = y_1'(0) = 1.$$

This procedure can have advantages. We may be genuinely interested in $y'(x)$. Mathematically speaking, having $y(x)$ gives us $y'(x)$ but computationally speaking, we may have an excellent approximation to $y(x)$ and no approximation to $y'(x)$ at all. When solved as a system both $y(x)$ and $y'(x)$ are treated the same and are computed equally well.

The standard initial value problem for an nth order equation is

$$y^{(n)} = f[x, y, y', \ldots, y^{(n-1)}],$$
$$y(a) = A_1,$$
$$y'(a) = A_2,$$
$$\vdots$$
$$y^{(n-1)}(a) = A_n.$$

The standard way to convert this to a system is to introduce the unknowns

$$y_1 = y,$$
$$y_2 = y',$$
$$\vdots$$
$$y_n = y^{(n-1)}$$

and then write

$$y_1' = y_2, \qquad\qquad y_1(a) = A_1,$$
$$y_2' = y_3, \qquad\qquad y_2(a) = A_2,$$
$$\vdots \qquad\qquad\qquad \vdots$$
$$y_n' = f(x, y_1, y_2, \ldots, y_n), \qquad y_n(a) = A_n.$$

The idea here is very broadly applicable. Just introduce new unknowns representing each derivative up to one less than the highest appearing for each unknown in the original problem. For example, suppose we encounter

$$u'' + u'v' = \sin x,$$
$$v' + v + u = \cos x.$$

Let

$$y_1 = u,$$
$$y_2 = u',$$
$$y_3 = v.$$

Then

$$y_1' = y_2,$$
$$y_2' = u'' = \sin x - y_2 y_3',$$
$$y_3' = \cos x - y_3 - y_1.$$

To put this in standard form we must eliminate y_3' from the second equation to arrive at

$$y_1' = y_2,$$
$$y_2' = \sin x - y_2(\cos x - y_3 - y_1),$$
$$y_3' = \cos x - y_3 - y_1.$$

Physical considerations may suggest other variables which are more appropriate. A common equation, one that occurs for example in the study of steady state diffusion, is of the form

$$\frac{d}{dx}\left(p(x)\frac{dy}{dx}\right) + q(x)y = g(x).$$

For positive differentiable $p(x)$ this can be converted to

$$y'' + \frac{p'(x)}{p(x)}y' + \frac{q(x)}{p(x)}y = \frac{g(x)}{p(x)}$$

which is put into standard form by setting $y_1 = y$ and $y_2 = y'$. In physical problems with layered media, the function $p(x)$ is discontinuous as we move from one medium into another. Our existence and uniqueness theorems apply only in separate pieces of the interval. Let us be specific; suppose that there are two layers, so that $p(x) \in C^1[a, c)$, $p(x) \in C^1(c, b]$, but $p(c-) \neq p(c+)$. [The standard notation here means

$$\lim_{\substack{x \to c \\ x < c}} p(x) = p(c-) \neq p(c+) = \lim_{\substack{x \to c \\ x > c}} p(x).]$$

Supposing that $q(x)$ and $g(x)$ are continuous on $[a, b]$ our theory says there is a unique solution with given initial values extending to the interface at c. If we define appropriate initial values at $c+$, the theory applies once again.

The typical physical condition is that the solution $y(x)$ and the flux $p(x)y'(x)$ are continuous. That is,

$$y(c+) = y(c-),$$

$$p(c+)y'(c+) = p(c-)y'(c-).$$

Notice that $y'(x)$ has a jump at c. Because of this condition the flux is a more natural variable than $y'(x)$. If we take $y_1 = y$, $y_2 = p(x)y'$, then

$$y_1' = y' = \frac{y_2}{p(x)},$$

$$y_2' = [p(x)y']' = -q(x)y_1 + g(x).$$

The interface condition in these variables merely requires us to compute approximations at c and to be careful in the evaluation of $p(x)$ there. Another advantage is that it is not necessary to use an analytical derivative for $p(x)$, as with the standard approach. This is fortunate since the derivative may be inconvenient (or impossible) to obtain.

This example brings to mind a point seen by the authors many times in computing practice. Often a scientist who has formulated a mathematical model as a system of differential equations in natural variables will eliminate them and combine the equations to obtain fewer equations but of higher order. This is done to facilitate analytical study. When it appears necessary to turn to numerical techniques, these higher order equations are converted back to a system, as required for most codes, by the standard device. In this artificial transformation the physical significance that the original variables had is lost and the resulting equations may prove less effective for the numerical solution.

To emphasize that there are many ways to obtain systems of equations let us examine an interesting special case. The equation is of the form

$$y'' + g(x)y' + h(x)y = 0 \tag{9}$$

and we wish to convert it to a system

$$y' = P(x)y + Q(x)z, \tag{10}$$

$$z' = R(x)y + S(x)z. \tag{11}$$

The functions y are to be the same and the choice of z, $P(x)$, $Q(x)$, $R(x)$, and $S(x)$ is at our disposal. We claim that the solution $y(x)$ of the system

(10) and (11) is the solution $y(x)$ of (9) provided that

$$P(x) + S(x) + \frac{Q'(x)}{Q(x)} + g(x) = 0, \tag{12}$$

$$R(x)Q(x) + P'(x) + P^2(x) + P(x)g(x) + h(x) = 0. \tag{13}$$

To see this, differentiate equation (10) to get

$$y'' = P'(x)y + P(x)y' + Q'(x)z + Q(x)z'$$

and substitute (11) to find

$$y'' = [P'(x) + Q(x)R(x)]y + P(x)y' + [Q'(x) + Q(x)S(x)]z.$$

Solve (10) for z and substitute the resulting expression in the last equation to find

$$y'' = \left(P'(x) + Q(x)R(x) - \frac{P(x)Q'(x)}{Q(x)} - P(x)S(x) \right) y$$

$$+ \left(P(x) + \frac{Q'(x)}{Q(x)} + S(x) \right) y'.$$

If we use (12) and (13), we see this is

$$y'' = -h(x)y - g(x)y',$$

as we desired. There are many choices for $P(x)$, $Q(x)$, $R(x)$, $S(x)$; we leave three as exercises.

EXERCISE 5

Verify that the following choices work:

a) $Q(x) \equiv 1, P(x) \equiv 0, R(x) = -h(x), S(x) = -g(x);$

b) $Q(x) = \frac{1}{E(x)}, P(x) \equiv 0, S(x) \equiv 0, R(x) = -h(x)E(x)$ where

$$E(x) = \exp\left(\int_a^x g(t)\, dt \right);$$

c) $Q(x) \equiv 1, P(x) = S(x) = \frac{-g(x)}{2}, R(x) = \frac{g'(x)}{2} + \frac{g^2(x)}{4} - h(x).$

Sometimes it is illuminating to study a problem by using several different transformations. An important part of the study of nonlinear mechanics is concerned with systems of the form

$$\frac{dx}{dt} = P(x, y),$$

$$\frac{dy}{dt} = Q(x, y).$$

Here the functions P and Q are to be smooth enough so that our existence and uniqueness theorems apply. The variables x and y are coordinates of position and t is time. Often one is concerned with the qualitative aspects of the motion and not the speed at which the path is traversed. By regarding y as a function of x so that

$$\frac{dy}{dx} = \frac{dy}{dt} \bigg/ \frac{dx}{dt} = \frac{Q(x, y)}{P(x, y)}$$

the role of time disappears. This analysis in the "phase plane" gives a great deal of insight but direct numerical solution of trajectories in the phase plane can lead to difficulties. The difficulties arise when either the denominator vanishes or the fraction becomes indeterminate. As an example of the latter situation consider

$$\frac{dx}{dt} = x,$$

$$\frac{dy}{dt} = y,$$

so that

$$\frac{dy}{dx} = \frac{y}{x}. \tag{14}$$

The first order equation is said to be singular at $x = 0$ since it does not satisfy the conditions of the existence theorem. The system is trivial to solve and leads to the conclusion that any problem consisting of (14) and $y(a) = A$ has the solution $y(x) = Ax/a$ as long as $a \neq 0$. Our theorems do apply directly to these problems if $a \neq 0$ and so we see there is only one solution then. If $a = 0$ and $A \neq 0$, there are no solutions at all but if $A = 0$, then $y(x) = cx$ is a solution for any constant c. This behavior on the part of the singular system looks bizarre but in suitable variables it is easily understood. Problems in which the derivative becomes infinite will be handled by a change of variables which is more appropriately discussed in Chapter 12.

EXERCISE 6

Write the following equations as systems of first order equations in standard form. Which of the original variables require initial values to specify an initial value problem when in standard form? (A common notation for differentiation with respect to time is $\dot{y} = dy/dt$.)

a) $y'' - (1 - y^2)y' + y = 0$

b) $x^3 \dfrac{d^3 y}{dx^3} + 6x^2 \dfrac{d^2 y}{dx^2} + 7x \dfrac{dy}{dx} = 0$

c) $x + \sin z = \dot{x}, \qquad \dot{z} + z = x$

d) $\ddot{y} + \dot{y} - \dot{v} = 0, \qquad y + \dot{v} + 3y + v = 0.$

A NUMERICAL METHOD

We shall discuss a simple numerical method in order to illustrate some of the considerations required in developing general purpose codes like those given in this text. This method itself is useful at those times when the codes we develop are inapplicable to a physical problem. Often, such problems can be solved after a little preparatory work using this simple numerical method. A number of examples illustrating this situation are examined in Chapter 12.

Suppose we wish to solve $y' = f(x, y)$, $y(a) = A$ on an interval $[a, b]$. This means that we want a table or graph of $y(x)$ over the interval. Notice that $y'(a) = f(a, y(a)) = f(a, A)$. Because of this the value

$$y_1 = y(a) + hy'(a) = A + hf(a, A)$$

approximates $y(a + h)$. The error is given by Taylor's theorem:

$$y(a + h) - y_1 = \frac{h^2}{2} y''(\xi)$$

$$= \frac{h^2}{2} y''(a) + \frac{h^3}{6} y'''(\zeta)$$

for some points ξ, ζ in $(a, a + h)$, provided that $y(x)$ is smooth enough. We say the approximation is $O(h^2)$ (in words, order of h^2). The order symbol "O" is used a great deal in succeeding chapters to indicate how well one expression approximates another. Typically, each expression will depend on the step size hence their difference will be a function of h. In the example, let $g(h) = y(a + h) - y_1$; then we say $g(h) = O(h^2)$. In general, when we write

$$g(h) = O(h^k),$$

we mean there is an $h_0 > 0$ and a constant $C \neq 0$ such that

$$|g(h)| \le Ch^k \qquad \text{for all } 0 < h \le h_0.$$

It is not necessary here to know a value for h_0 or C, just that they exist. In this particular instance we could take $h_0 = b - a$ and

$$C = \tfrac{1}{2} \max_{a \le x \le b} |y''(x)|.$$

The point of this is to describe *how fast* a quantity tends to zero as $h \to 0$. The following exercise develops some simple properties of the order relation which the reader must master if he is to understand the numerical solution of differential equations.

EXERCISE 7

Verify that if $g_1(h) = O(h^{k_1})$, $g_2(h) = O(h^{k_2})$, and α is any constant, then

$$\alpha g_1 = O(h^{k_1}),$$

$$g_1 g_2 = O(h^{k_1 + k_2}),$$

$$g_1 + g_2 = O(h^{k_3}), \qquad k_3 = \min(k_1, k_2).$$

We see that by making h sufficiently small we can obtain an approximate solution that is as accurate as we like. Ultimately we want to construct a table for the whole interval so (for the sake of efficiency) we want to use a relatively large h. To check the accuracy and adjust h we need to somehow estimate the error in y_1. Now

$$y''(x) = \frac{d}{dx} f(x, y(x)) = f_y(x, y(x))y'(x) + f_x(x, y(x))$$

$$= f_y(x, y(x))f(x, y(x)) + f_x(x, y(x)),$$

so

$$y''(a) = f_y(a, A)f(a, A) + f_x(a, A).$$

If we had analytical expressions for f_y and f_x, we could use them to evaluate $y''(a)$ and then estimate the error by using the expression:

$$\frac{h^2}{2} y''(a).$$

We say this estimate is asymptotically correct, meaning that as $h \to 0$ the ratio of the estimated error to the true error approaches unity. Put another way, the relative error of the estimate approaches zero as $h \to 0$. We can see that this is true for the relative error if the mean value theorem for derivatives is used:

$$\frac{\dfrac{h^2}{2} y''(\xi) - \dfrac{h^2}{2} y''(a)}{\dfrac{h^2}{2} y''(\xi)} = \frac{(\xi - a)y'''(\eta)}{y''(\xi)}.$$

In fact, since η and ξ lie in $(a, a + h)$ we see the relative error of the estimate is $O(h)$. [We assume the error $(h^2/2)y''(\xi)$ is not zero since, if this were the case, the relative error of its estimate would not be defined.]

To get more accurate approximations we might repeatedly differentiate the differential equation to obtain higher order derivatives and so use

$$y_1 = y(a) + hy'(a) + \cdots + \frac{h^k}{k!} y^{(k)}(a)$$

as an approximation to $y(a + h)$. This approximation is $O(h^{k+1})$ and the error can be estimated by

$$\frac{h^{k+1}}{(k + 1)!} y^{(k+1)}(a).$$

This approach to solving differential equations is called the Taylor series method.

What happens at the next step? Ideally we would solve

$$y' = f(x, y),$$

$$y(a + h) \qquad \text{given}$$

by taking a step from $a + h$. We do not yet have $y(a + h)$, rather we have only y_1. Thus the best we can do is solve

$$u' = f(x, u),$$

$$u(a + h) = y_1$$

for $u(x)$. This we do as before,

$$u_1 = u(a + h) + hu'(a + h) + \cdots$$

$$= y_1 + hf(a + h, y_1) + \cdots,$$

and $u_1 \doteq u(a + 2h)$. We can approximate $u(a + 2h)$ well for small h, but what about $y(a + 2h)$? We are seeing here the effects of errors propagating. At each step we can nearly obtain that solution of the differential equation with initial value given by the result of the previous step; we say the local error is small. The overall effect of these errors depends on the equation. If the problem is well posed and h is "small," then $y(a + 2h)$ is nearly the same as $u(a + 2h)$ since $y(x)$ and $u(x)$ are both solutions of the same differential equation with slightly different initial values at $a + h$. It is plausible, and true, that in taking $(b - a)/h$ steps, each with an error of order h^{k+1}, the worst error is of order h^k.

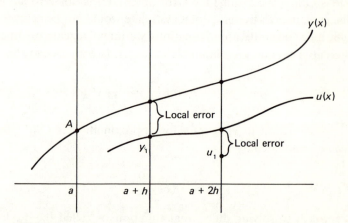

FIGURE 5

The local error of a step by step method for solving the initial value problem is the difference between the numerical approximation and the value of that solution curve of the differential equation going through the last approximation. How well the numerical approximation agrees with the solution of the problem posed depends on how the solution curves of the differential equation behave.

The determining factor in the propagation of error is the behavior of the solution curves of the differential equation. We attempt to follow a particular curve but, due to some error, may leave it and begin following a nearby curve in the family of solutions of the differential equation. If all the nearby solution curves tend to the one of interest as the integration proceeds, the effect of an error is damped out and we say the solution is stable. In the contrary case of the solution curves diverging from the one of interest, the effect of an error may grow as the integration progresses and we say the solution is unstable. As an illustration let us consider the solution $y(x) \equiv 0$ of $y' = \alpha y$. The general solution of the differential equation is $A \exp(\alpha x)$. If, say, $\alpha = -10^4$, all solutions converge rapidly to $y(x) \equiv 0$ so we find this solution is very stable. If $\alpha = 10^4$, the solution is very unstable. Theorem 2 gives a bound on how unstable a Lipschitzian problem can be but as the last example shows, the bound can be so large as to have no practical value. Suppose that we use some numerical method, such as the Taylor series, to solve $y' = 10^4 y$ with $y(0) = 0$. Suppose further that in the first step of length h we make a small local error of ε and on successive steps we make no error at all. Then as we saw above, after the first step we obtain exactly $u(x) = \varepsilon \exp(10^4(x - h))$ as an approximation to $y(x) \equiv 0$. And so even in these extraordinarily favorable circumstances there are serious limits on the accuracies which are attainable for large x. We shall prove that when using infinite precision arithmetic we can obtain numerical solutions to Lipschitzian problems as accurately as we like. In practice, however, step by step methods will yield satisfactory accuracies only for moderately stable problems.

The method of Taylor series is an excellent one, when it is feasible, but it does not have the broad applicability necessary for a general purpose code. Its basic idea is to approximate $y(x)$ near a by a polynomial

$$y(a) + y'(a)(x - a) + \frac{y''(a)}{2!}(x - a)^2 + \cdots + \frac{y^{(k)}(a)}{k!}(x - a)^k$$

which agrees with y and its first k derivatives at a. The scheme which is the subject of this text is based on a polynomial which agrees with y and y' at a and with y' at several other arguments near a. Since $y'(x) = f(x, y(x))$ is always readily available, this idea is not restricted by the necessity of other analytical differentiations, which is the obvious flaw of the Taylor series approach.

EXTENSION OF THE THEORY

Throughout the text we shall consider initial value problems (4) and (5) involving the class of functions $f(x, y)$ which are as smooth as required and which satisfy a Lipschitz condition for $a \leq x \leq b$, $-\infty < y < \infty$. The analysis is relatively simple with these assumptions and the results are easily seen to apply to a much larger class of problems. This is very important since most practical problems do not fall into the class. For one thing, they involve functions which may be smooth only in pieces. The basic theory can be applied to such problems if we just break up the interval of integration into corresponding pieces. While most real problems have partial derivatives, f_y, continuous and bounded in finite regions, they are usually not bounded for all y. As a prototype let us take $y' = y^2$, $y(0) = 1$ which we have already investigated. A way to treat such problems is to choose a constant M such that we will be uninterested in following the solution $y(x)$ if $|y(x)| > M$. This constant, M, could perhaps represent the overflow limit on our computer. Now let us consider the new problem

$$u' = f^*(x, u), \qquad u(a) = A$$

for which we define

$$f^*(x, u) = \begin{cases} f(x, M) & \text{if } u > M, \\ f(x, u) & \text{if } |u| \leq M, \\ f(x, -M) & \text{if } u < -M. \end{cases}$$

This new problem is Lipschitzian for all u. To verify this we first note that by assumption

$$\left| f^*(x, v) - f^*(x, w) \right| = \left| f(x, v) - f(x, w) \right| \leq L|v - w|$$

if both v and w are in $R = \{(x, u) \mid a \leq x \leq b, |u| \leq M\}$. If, for instance, $v > M$ and $|w| \leq M$, then

$$\left| f^*(x, v) - f^*(x, w) \right| = \left| f(x, M) - f(x, w) \right| \leq L|M - w| \leq L|v - w|.$$

Similar arguments work for all the remaining cases. The theory says there is a unique solution $u(x)$ to this initial value problem. But if $|u(x)| \leq M$ for $a \leq x \leq \beta$, then $u(x)$ satisfies

$$u(a) = A,$$

$$u'(x) = f(x, u(x)) \qquad \text{for} \quad a \leq x \leq \beta.$$

And so it is a solution of the original problem. By Theorem 2 it is also the only solution in this region. In actual computation no modification of the equation is necessary. For example, to solve the prototype just start integrating and when the numerical solution becomes too large, quit.

One can sharpen this modification procedure to consider a strip like $\{(x, y) \mid a \leq x \leq b, \; \alpha(x) < y < \beta(x)\}$ containing a solution $y(x)$ of interest, provided that $f(x, y)$ satisfies a Lipschitz condition there. The theory of the computational procedure will guarantee that for sufficiently small step sizes the numerical values are arbitrarily close to the desired solution; in particular, that they stay inside the strip. Of course, for large step sizes, the numerical values may fall outside the strip and bear no useful relation to the desired solution. Actually modifying the equation is sometimes necessary to avoid the annoyance of overflows on the computer or the embarrassment of the equation being undefined.

2

INTERPOLATION THEORY

The fundamental idea in this text is the approximation of a function $f(x)$ by a suitable interpolating polynomial $P(x)$, that is, a polynomial which agrees with f at several points x. Most introductory numerical analysis texts, e.g. [25], treat this subject, although often they do not develop the representations needed for solving differential equations. The reader familiar with both the Lagrange and Newton forms of the interpolating polynomial need only skim through this chapter. Because the subject is fundamental, a brief but complete development is given in this chapter.

In the preceding chapter we saw that expansion in a Taylor series led to a natural way of numerically approximating the solution of a differential equation. The Taylor series method is based on approximating a function $f(x)$ near a point x_0 by the Taylor polynomial

$$T_k(x) = f(x_0) + f'(x_0)(x - x_0) + \cdots + \frac{f^{(k-1)}(x_0)}{(k-1)!}(x - x_0)^{k-1}.$$

This polynomial agrees with f and its first $k - 1$ derivatives at x_0,

$$T_k^{(j)}(x_0) = f^{(j)}(x_0) \qquad j = 0, 1, \ldots, k - 1,$$

or, as we shall say, interpolates to these values at x_0. For reasons discussed in Chapter 1 it is more practical to obtain values of f at several points than to obtain the values of several derivatives of f at one point. A polynomial $P_k(x)$ of degree $k - 1$ will be constructed which interpolates to $f(x)$ at k distinct points $x_0, x_1, \ldots, x_{k-1}$:

$$P_k(x_i) = f(x_i) \qquad i = 0, 1, \ldots, k - 1. \tag{1}$$

There is one, and only one, polynomial of degree $k - 1$ which satisfies these interpolation conditions; but, it may be written in several different ways. Newton's form of the interpolating polynomial is analogous to the Taylor polynomial, and we shall use it to solve differential equations in a way quite similar to the Taylor series method. Because of this, and because the reader is probably familiar with Taylor expansions from a study of the calculus, we shall emphasize the analogy as we develop the interpolation method. We shall also develop Lagrange's form of the interpolating polynomial, since it is often more convenient for analytical purposes than Newton's form. For example, the existence of $P_k(x)$ is obvious in Lagrange's form.

A few comments about notation and terminology will be helpful. We shall often write f_i for $f(x_i)$. When we describe a polynomial as being of degree $k - 1$ we do not exclude the possibility that it is of a lower degree. For example, we might say

$$3 + 2x + x^2 + 0x^3$$

is of degree 3. The more precise term "of exact degree $k - 1$" means that x^{k-1} is the highest power with a non-zero coefficient, the example being of exact degree 2. We write $P_k(x)$ for the polynomial of degree $k - 1$ satisfying (1). Thus the subscript k is chosen to agree with the number of interpolation points rather than with the degree of the polynomial. This is more consistent with our subsequent applications to differential equations and is more convenient for the computer implementations.

Lagrange constructed a polynomial $P_k(x)$ having the properties of equation (1) from a set of fundamental polynomials $l_i(x)$, each of degree $k - 1$ and each satisfying

$$\begin{aligned} l_i(x_j) &= 1 \qquad \text{if} \quad j = i, \\ &= 0 \qquad \text{if} \quad j \neq i. \end{aligned} \tag{2}$$

Clearly

$$l_i(x) = \frac{(x - x_0)(x - x_1)\ldots(x - x_{i-1})(x - x_{i+1})\ldots(x - x_{k-1})}{(x_i - x_0)(x_i - x_1)\ldots(x_i - x_{i-1})(x_i - x_{i+1})\ldots(x_i - x_{k-1})}$$

is of degree $k - 1$ and satisfies the equations of (2). Then

$$P_k(x) = f_0 l_0(x) + f_1 l_1(x) + \cdots + f_{k-1} l_{k-1}(x) \tag{3}$$

is an interpolating polynomial since $P_k(x)$ is of the correct degree and

$$P_k(x_i) = f_0 l_0(x_i) + \cdots + f_i l_i(x_i) + \cdots + f_{k-1} l_{k-1}(x_i)$$
$$= f_i \quad \text{for each } x_i.$$

Equation (3) is called Lagrange's form of the interpolating polynomial which can be written in the more compact notation

$$P_k(x) = \sum_{i=0}^{k-1} f_i \prod_{\substack{j=0 \\ j \neq i}}^{k-1} \left(\frac{x - x_j}{x_i - x_j} \right).$$

This argument shows that there is at least one polynomial of degree $k - 1$ which will assume given values at each of k distinct points. We shall now prove there is only one interpolating polynomial of this degree.

THEOREM 1   Let $x_0, x_1, \ldots, x_{k-1}$ be k distinct real points and let $f_0, f_1, \ldots, f_{k-1}$ be corresponding function values. Then there exists one, and only one, polynomial $P_k(x)$ of degree $k - 1$ satisfying

$$P_k(x_i) = f_i^- \quad i = 0, 1, \ldots, k - 1.$$

Proof.   We have already seen that one interpolating polynomial $P(x)$ of degree $k - 1$ exists. Let $\bar{P}(x)$ be another interpolating polynomial of degree $k - 1$. Then the polynomial $D(x) = P(x) - \bar{P}(x)$ is of degree $k - 1$ and has k real zeros $x_0, x_1, \ldots, x_{k-1}$; more precisely, it is of the exact degree $m \leq k - 1$ so

$$D(x) = a_m x^m + a_{m-1} x^{m-1} + \cdots + a_0$$

where $a_m \neq 0$. From the calculus, Rolle's theorem states that the derivative of a function vanishes at least once between every pair of real zeros of the function. Consequently, the polynomial

$$D'(x) = m a_m x^{m-1} + \text{lower degree terms}$$

has at least $k - 1$ zeros. Repeating this argument we conclude

$$D^{(m)}(x) = m! a_m$$

has at least $k - m$ zeros. But this is impossible, since $m! a_m \neq 0$, and this contradiction establishes the theorem.

The uniqueness of the interpolating polynomial means that if two polynomials interpolate to the same data and there is sufficient data, they are merely different representations of the same polynomial. This conclusion has two important implications. First, the order in which data are arranged is immaterial. Second, a numerical method based on interpolation can be analyzed using one form of the interpolating polynomial and can be applied computationally using another form.

In solving differential equations, one is continually going from an interpolation on one set of data points to an interpolation on another set obtained by dropping a point, adding a point, or both. Each time the data set is changed Lagrange's form must be entirely recomputed. Economies can be realized by using a form which takes advantage of the fact that most of the data stays the same. The Taylor polynomials suggest that this is possible. To go from interpolating $f(x_0), \ldots, f^{(k-1)}(x_0)$ to interpolating $f(x_0), \ldots, f^{(k-1)}(x_0), f^{(k)}(x_0)$ we have

$$T_{k+1}(x) = T_k(x) + \frac{f^{(k)}(x_0)}{k!}(x - x_0)^k.$$

To attempt something equally simple in the case of increasing the data set from k to $k + 1$ points we shall require that

$$P_{k+1}(x) = P_k(x) + R_{k+1}(x)$$

where $R_{k+1}(x)$ must be of degree k. That is, adding a data point will be handled by adding a correction term to the current interpolating polynomial. To see how this might work, suppose x_0 is the only data point; clearly

$$P_1(x) = f_0$$

interpolates at x_0. If the set is increased to $\{x_0, x_1\}$, then for

$$P_2(x) = P_1(x) + R_2(x)$$

it is required that

$$R_2(x) = (x - x_0)a_1$$

in order for $P_2(x)$ to interpolate to f_0. The second interpolation condition determines a_1:

$$f_1 = P_2(x_1) = f_0 + (x_1 - x_0)a_1$$

or

$$a_1 = \frac{f_0 - f_1}{x_0 - x_1}.$$

The general scheme is easy to see. We claim that

$$R_{k+1}(x) = (x - x_0)(x - x_1)\ldots(x - x_{k-1})a_k$$

for a suitable constant a_k. The interpolation conditions on the old data are satisfied by the form of $R_{k+1}(x)$:

$$P_{k+1}(x_i) = P_k(x_i) + R_{k+1}(x_i),$$
$$= f_i + 0, \qquad i = 0, 1, \ldots, k - 1.$$

The condition at x_k requires that

$$P_{k+1}(x_k) = f_k = P_k(x_k) + a_k \prod_{j=0}^{k-1} (x_k - x_j)$$

or

$$a_k = \frac{(f_k - P_k(x_k))}{\displaystyle\prod_{j=0}^{k-1} (x_k - x_j)}.$$

Because x_k is distinct from the other data points, a_k is well defined. Since all the interpolation conditions are satisfied by $P_k(x) + R_{k+1}(x)$ and it is a polynomial of degree k, it must also be $P_{k+1}(x)$.

The coefficient a_k is an example of a divided difference of order k and will be written in the standard notation $f[x_0, x_1, \ldots, x_k]$ to display the data it depends on. As a special case, $a_0 = f_0 = f[x_0]$. In this notation

$$P_k(x) = f[x_0] + (x - x_0)f[x_0, x_1] + \cdots$$
$$+ (x - x_0)(x - x_1)\ldots(x - x_{k-2})f[x_0, x_1, \ldots, x_{k-1}]. \qquad (4)$$

This is the Newton or divided difference form of the interpolating polynomial.

The relation between Newton's form of the interpolating polynomial and the Taylor polynomial is easy to see for the simple case of $P_1(x)$. Using the mean value theorem for derivatives, we find that $f[x_0, x_1] = f'(\xi)$ for some point ξ between x_0 and x_1 so,

$$P_1(x) = f(x_0) + (x - x_0)f'(\xi).$$

If we let $x_1 \to x_0$, the polynomial $P_1(x)$ goes into $T_1(x)$. We shall prove similar results for the general case, but first we must develop some properties of divided differences and discuss how to compute them.

It will prove useful to have an explicit expression for the dependence of divided differences on the function values. From what we have just seen

$$P_k(x) = x^{k-1} f[x_0, x_1, \ldots, x_{k-1}] + \text{lower degree terms.}$$

The Lagrangian form has

$$P_k(x) = \sum_{i=0}^{k-1} f_i \prod_{\substack{j=0 \\ j \neq i}}^{k-1} \left(\frac{x - x_j}{x_i - x_j} \right)$$

$$= x^{k-1} \sum_{i=0}^{k-1} f_i \prod_{\substack{j=0 \\ j \neq i}}^{k-1} \frac{1}{x_i - x_j} + \text{lower degree terms.}$$

From the uniqueness theorem, these are different forms of the same polynomial; in particular, the coefficients of x^{k-1} must be the same:

$$f[x_0, x_1, \ldots, x_{k-1}] = \sum_{i=0}^{k-1} f_i \prod_{\substack{j=0 \\ j \neq i}}^{k-1} \frac{1}{x_i - x_j}. \tag{5}$$

This explicit representation of the divided difference makes it clear that the order in which data points are used does not affect the value of the divided difference, since the only effect is on the order of terms in the sum. This important fact will be used often.

With (5) we can prove a simple relation which will provide an efficient way to compute divided differences and interpolating polynomials:

$$f[x_0, x_1, \ldots, x_{k-1}] = \frac{f[x_0, x_1, \ldots, x_{k-2}] - f[x_1, x_2, \ldots, x_{k-1}]}{x_0 - x_{k-1}}. \tag{6}$$

To see this, observe that

$$f[x_0, x_1, \ldots, x_{k-2}] - f[x_1, x_2, \ldots, x_{k-1}]$$

$$= f_0 \prod_{j=1}^{k-2} \frac{1}{x_0 - x_j} + \sum_{i=1}^{k-2} f_i \prod_{\substack{j=0 \\ j \neq i}}^{k-2} \frac{1}{x_i - x_j} - \sum_{i=1}^{k-2} f_i \prod_{\substack{j=1 \\ j \neq i}}^{k-1} \frac{1}{x_i - x_j}$$

$$- f_{k-1} \prod_{j=1}^{k-2} \frac{1}{x_{k-1} - x_j}$$

$$= f_0 \prod_{j=1}^{k-2} \frac{1}{x_0 - x_j} + \sum_{i=1}^{k-2} f_i \left[\frac{1}{x_i - x_0} - \frac{1}{x_i - x_{k-1}} \right] \prod_{\substack{j=1 \\ j \neq i}}^{k-2} \frac{1}{x_i - x_j}$$

$$- f_{k-1} \prod_{j=1}^{k-2} \frac{1}{x_{k-1} - x_j}.$$

Dividing by $x_0 - x_{k-1}$ yields, on the right hand side,

$$f_0 \prod_{j=1}^{k-1} \frac{1}{x_0 - x_j} + \sum_{i=1}^{k-2} f_i \prod_{\substack{j=0 \\ j \neq i}}^{k-1} \frac{1}{x_i - x_j} + f_{k-1} \prod_{j=0}^{k-2} \frac{1}{x_{k-1} - x_j}$$

which, by comparison with (5), proves equation (6). This latter form can be used to generate the differences needed in the Newton form as indicated by the lines in the following table.

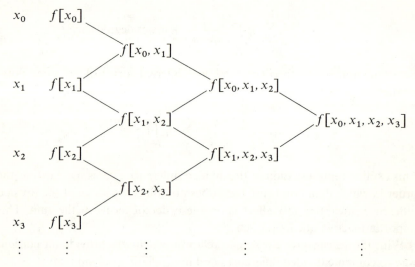

This table actually provides the coefficients needed for a great many interpolating polynomials, depending on which points one uses for interpolation. In particular, another form of $P_k(x)$ is found by considering it as interpolating on the data points in reversed order $x_{k-1}, x_{k-2}, \ldots, x_0$:

$$P_k(x) = f[x_{k-1}] + (x - x_{k-1})f[x_{k-1}, x_{k-2}] + \cdots$$
$$+ (x - x_{k-1})(x - x_{k-2})\ldots(x - x_1)f[x_{k-1}, \ldots, x_0].$$

Let us now look at a few simple examples which illustrate the mechanics of interpolation. First let

x_i	f_i
1	3
3	5
4	4

be data through which we pass a quadratic polynomial. In Lagrange's form the fundamental polynomials are:

$$l_0(x) = \frac{x-3}{1-3} \cdot \frac{x-4}{1-4},$$

$$l_1(x) = \frac{x-1}{3-1} \cdot \frac{x-4}{3-4},$$

$$l_2(x) = \frac{x-1}{4-1} \cdot \frac{x-3}{4-3},$$

and

$$P_3(x) = 3l_0(x) + 5l_1(x) + 4l_2(x). \tag{7}$$

The difference form of the polynomial requires constructing a table of divided differences:

| x_i | $f[x_i]$ | $f[x_i, x_{i+1}]$ | $f[x_i, x_{i+1}, x_{i+2}]$ |

$$
\begin{array}{cccc}
1 & 3 \\
 & & 1 \\
3 & 5 & & -2/3 \\
 & & -1 \\
4 & 4 \\
\end{array}
$$

Once it is constructed we immediately see that

$$P_3(x) = 3 + (x-1)(1) + (x-1)(x-3)(-\tfrac{2}{3}), \tag{8a}$$

since the coefficients of $P_3(x)$ are the entries along the upper diagonal. Likewise, if we consider the data in reversed order we have

$$\bar{P}_3(x) = 4 + (x-4)(-1) + (x-4)(x-3)(-\tfrac{2}{3}) \tag{8b}$$

where now the coefficients lie along the lower diagonal.

EXERCISE 1

Verify that the polynomials in (7) and (8) are algebraically equivalent.

Suppose that the point corresponding to $x_3 = 7$, $f_3 = 9$ is added; we merely augment the existing table to include this information:

x_i	$f[x_i]$	$f[x_i, x_{i+1}]$	$f[x_i, x_{i+1}, x_{i+2}]$	$f[x_i, x_{i+1}, x_{i+2}, x_{i+3}]$

$$
\begin{array}{ccccc}
x_i & f[x_i] & & & \\
1 & 3 & & & \\
 & & 1 & & \\
3 & 5 & & -2/3 & \\
 & & -1 & & 2/9 \\
4 & 4 & & 2/3 & \\
 & & 5/3 & & \\
7 & 9 & & &
\end{array}
$$

Then

$$P_4(x) = 3 + (x - 1)(1) + (x - 1)(x - 3)(-\tfrac{2}{3}) + (x - 1)(x - 3)(x - 4)(\tfrac{2}{9})$$
$$= P_3(x) + (x - 1)(x - 3)(x - 4)f[1, 3, 4, 7].$$

Likewise, if we wish to delete the pair $(4, 4)$ from the original table, we simply drop the entire lower diagonal to obtain the polynomial interpolating to the remaining data:

$$P_2(x) = 3 + (x - 1)(1)$$
$$= P_3(x) - (x - 1)(x - 3)f[1, 3, 4].$$

EXERCISE 2

Suppose that the point $(0, 1)$ is added to the original table. Express $\bar{P}_4(x)$ in terms of $\bar{P}_3(x)$.

Interpolation by divided differences is simplified when the spacing between data points is constant. This is very important in the numerical solution of differential equations. Although it is necessary to allow unequal spacing, it is usually true that the spacing is constant. We want to take advantage of any economies offered by this situation. Furthermore some parts of the theory are rigorously true only when the spacing is constant, so we must fully understand the special case to gain insight for the heuristics essential to an understanding of the general case.

When the spacing is a constant $h = x_i - x_{i-1}$ for all i, it is more convenient to use backward differences than divided differences. These backward differences are simply scaled divided differences which remove the dependence on the spacing h. The backward difference operators ∇^i are defined by

$$\nabla^0 f_k = f_k,$$
$$\nabla^1 f_k = \nabla f_k = f_k - f_{k-1}, \tag{9}$$
$$\vdots \qquad \vdots \qquad \vdots$$
$$\nabla^i f_k = \nabla(\nabla^{i-1} f_k) = \nabla^{i-1} f_k - \nabla^{i-1} f_{k-1}.$$

The name "backward differences" is very natural in view of this definition. We shall show that

$$\nabla^i f_k = i! h^i f[x_k, x_{k-1}, \ldots, x_{k-i}]. \tag{10}$$

The relation (10) is obvious for $i = 0$ and 1 from the definitions

$$\nabla^0 f_k = f_k = 0! h^0 f[x_k]$$

and

$$\nabla^1 f_k = f_k - f_{k-1} = 1! \, h^1 f[x_k, x_{k-1}].$$

Suppose that it is true for $i = m$. Then

$$\nabla^{m+1} f_k = \nabla^m f_k - \nabla^m f_{k-1}$$
$$= m! h^m f[x_k, x_{k-1}, \ldots, x_{k-m}] - m! h^m f[x_{k-1}, x_{k-2}, \ldots, x_{k-m-1}]$$
$$= m! h^m (m+1) h \frac{f[x_k, \ldots, x_{k-m}] - f[x_{k-1}, \ldots, x_{k-m-1}]}{x_k - x_{k-m-1}}$$
$$= (m+1)! h^{m+1} f[x_k, x_{k-1}, \ldots, x_{k-m-1}]$$

on using the relation (6) and the fact that the value of divided differences does not depend on the order of the data points. By induction the relation (10) holds for all i. By considering the expression (5) in the special case of constant spacing we can find the dependence of the backward differences on the function values. It is stated neatly in terms of binomial coefficients $\binom{i}{m}$ which are defined as

$$\binom{i}{m} = \frac{i!}{m!(i-m)!}.$$

The relation is

$$\nabla^i f_k = \sum_{m=0}^{i} (-1)^m f_{k-m} \binom{i}{m}. \tag{11}$$

EXERCISE 3

Prove (11).

In the notation of backward differences the interpolating polynomial for equally spaced data points $x_{k-1}, x_{k-2}, \ldots, x_0$ is written

$$P_k(x) = f_{k-1} + (x - x_{k-1})\frac{\nabla f_{k-1}}{h} + (x - x_{k-1})(x - x_{k-2})\frac{\nabla^2 f_{k-1}}{2h^2}$$

$$+ \cdots + (x - x_{k-1})(x - x_{k-2})\cdots(x - x_1)\frac{\nabla^{k-1} f_{k-1}}{(k-1)!\, h^{k-1}}.$$

This polynomial is obtained just as when using divided differences but backward differences are easier to compute. As an example, consider the data below.

x_i	f_i
1	2
3	6
5	−1

The table of differences is

$\nabla^0 f_i$	$\nabla^1 f_i$	$\nabla^2 f_i$
2		
	4	
6		−11
	−7	
−1		

and the interpolating polynomial is

$$P_3(x) = -1 + (x - 5)\frac{(-7)}{(2)} + (x - 5)(x - 3)\frac{(-11)}{2!(2)^2}.$$

Taylor's theorem gives an error expression for approximation by the Taylor polynomial. If f has k continuous derivatives on an interval $[a,b]$ and $T_k(x)$ is the Taylor polynomial interpolating to $f(x_0), f'(x_0), \ldots, f^{(k-1)}(x_0)$ at some point x_0 in $[a,b]$, then for each x in $[a,b]$

$$f(x) - T_k(x) = \frac{f^{(k)}(\xi(x))}{k!}(x - x_0)^k$$

for some ξ between x_0 and x. We notice that

$$T_{k+1}(x) - T_k(x) = \frac{f^{(k)}(x_0)}{k!}(x - x_0)^k$$

approximates the error when x is close to x_0. Indeed, if $f^{(k)}(x_0) \neq 0$, then this estimate is asymptotically correct as x tends to x_0; recall the arguments along this line at the end of Chapter 1. This is a very convenient estimate of the error since if the error is not sufficiently small, we have a better approximation, $T_{k+1}(x)$, immediately available.

We shall find that results exactly analogous to those for the Taylor polynomial are also true for the interpolating polynomial but first we must derive an error expression.

THEOREM   2   Let $f(x)$ have k continuous derivatives on an interval $[a, b]$. Let $x_0, x_1, \ldots, x_{k-1}$ be distinct points in $[a, b]$ and let $P_k(x)$ be the polynomial of degree $k - 1$ interpolating to the corresponding function values $f_0, f_1, \ldots, f_{k-1}$. Let $\omega(x) = \prod_{i=0}^{k-1} (x - x_i)$. Then for each x in $[a, b]$

$$f(x) - P_k(x) = E(x)\omega(x),$$

where $E(x)$ is a continuous function which has the representations

$$E(x) = \frac{f'(x_i) - P'(x_i)}{\omega'(x_i)} \qquad \text{if } x = x_i,$$

$$= \frac{f^{(k)}(\xi(x))}{k!}. \qquad \text{if } x \neq x_i.$$

The point $\xi(x)$ is an unknown point, depending on x, which lies in the smallest interval containing x and the data points.

Proof.   Obviously if x is not one of the x_i,

$$E(x) = \frac{f(x) - P_k(x)}{\omega(x)}.$$

Since the numerator is continuous, as is the denominator, $E(x)$ is also continuous. As x approaches some x_i, the numerator and denominator vanish simultaneously. However, since the points of interpolation are distinct, $\omega'(x_i) \neq 0$ and L'Hospital's rule says that

$$E(x_i) = \frac{f'(x_i) - P_k'(x_i)}{\omega'(x_i)}.$$

This defines $E(x)$ as a continuous function for all x in $[a, b]$ and proves the expression stated in the theorem for points of interpolation.

For x not a point of interpolation we argue as follows. Let x be a *fixed* point and define the function of t

$$F(t) = f(t) - P_k(t) - E(x)\omega(t).$$

Note that $E(x)$ is a constant here. By its construction $F(t)$ has k continuous derivatives and $k + 1$ zeros. The zeros are the k interpolation points and x, by the definition of $E(x)$. Using Rolle's theorem we find that $F'(t)$ has at least k zeros in $[a, b]$, $F''(t)$ has at least $k - 1$ zeros, and so on. Finally, $F^{(k)}(t)$ has at least one zero which we shall call $\xi(x)$. Then

$$0 = F^{(k)}(\xi(x)) = f^{(k)}(\xi(x)) - E(x) \cdot k!,$$

hence

$$E(x) = \frac{f^{(k)}(\xi(x))}{k!},$$

and the proof is complete.

This theorem provides a better understanding of divided differences as well as practical estimates of the error of interpolation. Let x be a fixed data point distinct from $x_0, x_1, \ldots, x_{k-1}$ and let $P_{k+1}(t)$ interpolate to $f(t)$ at these $k+1$ points. The extra data point is incorporated by

$$P_{k+1}(t) = P_k(t) + (t - x_0)(t - x_1)\cdots(t - x_{k-1}) f[x_0, \ldots, x_{k-1}, x].$$

The interpolation condition at x implies

$$f(x) - P_k(x) = P_{k+1}(x) - P_k(x)$$
$$= (x - x_0)(x - x_1)\cdots(x - x_{k-1})f[x_0,\ldots,x_{k-1},x].$$

Comparing this to the conclusion of the theorem, we find that

$$f[x_0, x_1,\ldots,x_{k-1}, x] = \frac{f^{(k)}(\xi(x))}{k!}$$

for some $\xi(x)$ in the smallest interval containing x and the $x_0, x_1,\ldots,x_{k-1}$. We usually denote $k+1$ data points as $x_0, x_1,\ldots,x_k$ so let us restate the conclusion in the customary form:

$$f[x_0, x_1,\ldots,x_k] = \frac{f^{(k)}(\xi)}{k!}. \tag{12}$$

Using this representation of a divided difference we see that if $f(x)$ is sufficiently smooth, there are points $\xi_1, \xi_2,\ldots,\xi_{k-1}$ lying in the smallest interval containing the k data points $x_0, x_1,\ldots,x_{k-1}$ such that the divided difference form of $P_k(x)$ given by equation (4) can be written

$$P_k(x) = f(x_0) + f'(\xi_1)(x - x_0) + \frac{f''(\xi_2)}{2}(x - x_0)(x - x_1)$$

$$+ \cdots + \frac{f^{(k-1)}(\xi_{k-1})}{(k-1)!}(x - x_0)(x - x_1)\cdots(x - x_{k-2}).$$

In this form the analogy to the Taylor polynomial is very evident. If x_i tends to x_0 for $i = 1, 2,\ldots,k - 1$, we see that $P_k(x)$ tends to $T_k(x)$. For data points which are "close" together there is little difference between the two polynomials.

Just as with the Taylor polynomials, we can easily estimate the error of interpolation by

$$f(x) - P_k(x) \doteq (x - x_0)(x - x_1)\cdots(x - x_{k-1})f[x_0,\ldots,x_{k-1}, x_k]$$
$$= P_{k+1}(x) - P_k(x).$$

If $f(x)$ has a continuous derivative of order k, this estimate is

$$(x - x_0)(x - x_1)\cdots(x - x_{k-1})\frac{f^{(k)}(\eta)}{k!}$$

and the true error is

$$(x - x_0)(x - x_1)\cdots(x - x_{k-1})\frac{f^{(k)}(\xi)}{k!}$$

where η and ξ both lie in the smallest interval containing x and the data points $x_0, x_1, \ldots, x_k$. How good the practical estimate is depends on how much $f^{(k)}$ varies over this interval. If $f^{(k)}(x_0) \neq 0$, then the estimate is asymptotically correct, for it is easy to verify that the ratio of the estimated and true errors tends to unity as the interval collapses to the point x_0.

When there is enough data to be flexible about the degree of the inter-polating polynomial, these ideas can be used to determine the "right" degree. The representation of the divided difference as a scaled derivative (equation 12) makes it easy to see what a difference table for a polynomial is like. If $f(x)$ is a polynomial of degree m, all divided differences of order m have the same constant nonzero value and all higher order differences are zero. When $f(x)$ is a general function, the degree is chosen so that this behavior is approximately true. Of course whether these quantities can be regarded as zero depends on the accuracy wanted. In solving differential equations one approximates an $f(x)$ first using a set of data points $\{x_j, x_{j+1}, \ldots, x_{j+k}\}$ and then shifting to a new set of data points $\{x_{j+1}, x_{j+2}, \ldots, x_{j+k}, x_{j+k+1}\}$. The spacing of the new data point is manipulated so that for this set of data, approximation by a polynomial is reasonable. The order of interpolation can be varied by dropping extra points, for example using $\{x_{j+2}, \ldots, x_{j+k+1}\}$, or retaining an extra point and using $\{x_j, \ldots, x_{j+k+1}\}$. The nature of this situation is such that an appropriate degree for interpolation changes infrequently, and then ordinarily only by one. As applied to differential equations it is possible to do very well in choosing the "right" degree.

3

ADAMS FORMULAS

The methods to be developed for approximating the solution of

$$y'(x) = f(x, y(x)),$$ (1)

$$y(a) = A$$ (2)

on an interval $[a, b]$ do so on a mesh, a set of points $\{x_0, x_1, \ldots\}$ in $[a, b]$. Frequently these points are equally spaced with a mesh spacing, or step size, h:

$$x_n = a + nh \qquad n = 0, 1, \ldots.$$

The general case allows mesh points separated by unequal step sizes $h_1, h_2, \ldots$ so that

$$x_0 = a,$$

$$x_n = x_{n-1} + h_n \qquad n = 1, 2, \ldots.$$

When using the codes of Chapter 10, one usually finds that the mesh consists of sequences of equally spaced mesh points with transitions of a few unequally spaced points. The notation y_n will always mean an approximation to the solution $y(x)$ of equations (1) and (2) at the mesh point x_n:

$$y_n \doteq y(x_n).$$

Because $y(x)$ satisfies (1), an approximation of $y(x_n)$ leads to an approximation of $y'(x_n)$, namely:

$$f_n = f(x_n, y_n) \doteq y'(x_n) = f(x_n, y(x_n)).$$

We use this approximation throughout the text. The basic computational
task is to advance the numerical solution to x_{n+1} after having computed
$y_0, y_1, \ldots, y_n$.

Any solution of (1) can be written as

$$y(x_{n+1}) = y(x_n) + \int_{x_n}^{x_{n+1}} y'(t)\,dt = y(x_n) + \int_{x_n}^{x_{n+1}} f(t, y(t))\,dt.$$

The Adams methods approximate this solution by replacing $f(t, y(t))$ with
a polynomial interpolating to computed derivative values, f_i, and then
integrating the polynomial. These methods can be as effective as any in use
today and have the advantage of being sufficiently simple in form that
they are relatively well understood.

The Adams-Bashforth formula of order k at x_n uses a polynomial
$P_{k,n}(x)$ interpolating the computed derivatives at the k preceding points,

$$P_{k,n}(x_{n+1-j}) = f_{n+1-j} \qquad j = 1, 2, \ldots, k.$$

These derivatives and y_n must be stored from the preceding step. An
approximation to the solution at x_{n+1} is obtained from

$$y_{n+1} = y_n + \int_{x_n}^{x_{n+1}} P_{k,n}(t)\,dt. \tag{3}$$

More generally, an approximation can be obtained for all x near x_n by
using

$$y(x) \doteq y_n + \int_{x_n}^{x} P_{k,n}(t)\,dt.$$

The analysis can be made wholly in terms of the values y_n at the mesh points
x_n because they determine the approximation everywhere. In Chapter 4, we
prove that the approximation between mesh points is as good as at the mesh
points themselves. This fact is heavily used in computing practice.

As discussed in Chapter 2, there are several ways of representing the
interpolating polynomial, so that equation (3) has several formulations. We
develop first the Lagrangian, or ordinate, form which is both simple and

particularly suited to discussing how well y_{n+1} approximates $y(x_{n+1})$. Lagrange's form of the interpolating polynomial is

$$P_{k,n}(x) = \sum_{i=1}^{k} l_i(x) f_{n+1-i}$$

where

$$l_i(x) = \prod_{\substack{j=1 \\ j \neq i}}^{k} \frac{x - x_{n+1-j}}{x_{n+1-i} - x_{n+1-j}} \qquad i = 1, \dots, k.$$

Substituting into equation (3), we find

$$y_{n+1} = y_n + \sum_{i=1}^{k} f_{n+1-i} \int_{x_n}^{x_{n+1}} l_i(t) \, dt.$$

The Adams-Bashforth formula is usually written

$$y_{n+1} = y_n + h_{n+1} \sum_{i=1}^{k} \alpha_{k,i} f_{n+1-i}, \qquad (4)$$

where

$$\alpha_{k,i} = \frac{1}{h_{n+1}} \int_{x_n}^{x_{n+1}} l_i(t) \, dt;$$

or, in terms of the variable $s = (t - x_n)/h_{n+1}$,

$$\alpha_{k,i} = \int_0^1 l_i(x_n + s h_{n+1}) \, ds.$$

The classical formulation considers only mesh points equally spaced with a constant step size h. Introducing the $\alpha_{k,i}$ is then computationally convenient because they do not depend on h. In the general case the $\alpha_{k,i}$ depend only on the relative spacing of the mesh points and not on the scale of the independent variable.

EXERCISE 1

If the independent variable is changed by a scale factor σ so that x_{n+1-i} is replaced by σx_{n+1-i} for each i, show that the $\alpha_{k,i}$ are unaffected. What happens to h_{n+1} under this scale change? When the step size is a constant h, show that the $\alpha_{k,i}$ do not depend on the step size.

There is an equivalent Adams-Bashforth formula based on the divided difference form of the interpolating polynomial. The case of general mesh spacing will be treated in detail in Chapter 5 with careful attention to computational matters. For now, however, we shall just look at the classical formulation for constant step size in terms of backward differences. In this case

$$P_{k,n}(x) = f_n + \frac{(x - x_n)}{h}\nabla f_n + \cdots + \frac{(x - x_n)(x - x_{n-1})\cdots(x - x_{n+2-k})}{h^{k-1}(k-1)!}\nabla^{k-1}f_n.$$

The classical Adams-Bashforth formula in backward difference form is found from (3) to be

$$y_{n+1} = y_n + h\sum_{i=1}^{k} \gamma_{i-1}\nabla^{i-1}f_n \tag{5}$$

where again we use $s = (t - x_n)/h_{n+1}$ to see

$$\gamma_0 = 1,$$

$$\gamma_i = \frac{1}{i!\,h}\int_{x_n}^{x_{n+1}} \frac{(t - x_n)(t - x_{n-1})\cdots(t - x_{n+1-i})}{h^{i-1}}\,dt \tag{6}$$

$$= \frac{1}{i!}\int_0^1 s(s+1)\cdots(s+i-1)\,ds \qquad i \ge 1.$$

An important advantage of the difference form of the interpolating polynomial is that raising or lowering the degree amounts to adding or dropping a term. From the definition of y_{n+1} in equation (3) it can be seen that the same is true of changing the order of the Adams-Bashforth formula when divided differences are used. The fact that the γ_i in (5) are independent of k is an illustration of this basic fact. In contrast the Lagrangian form changes all the $\alpha_{k,i}$ when k is changed.

Let us consider a simple case. For $k = 1$, $P_{1,n}(x)$ is the constant f_n, so from equation (3)

$$y_{n+1} = y_n + \int_{x_n}^{x_{n+1}} f_n\,dt$$

$$= y_n + hf_n.$$

This simple method, which we also saw in connection with Taylor series, is called Euler's method. When $n = 0$, the above formula is $y_1 = y_0 + hf_0$. What are y_0 and f_0 here? The obvious choice for y_0 is $y_0 = A$ since

$y_0 \doteq y(x_0) = A$. Then we use $f_0 = f(x_0, y_0)$, as always. The first order Adams-Bashforth formula is "self-starting" because it requires only information available from the problem itself. On the other hand, for $k = 2$ and general mesh spacing, we see

$$\alpha_{2,1} = \int_0^1 \frac{x_n + sh_{n+1} - x_{n-1}}{x_n - x_{n-1}} ds = 1 + \frac{1}{2} \frac{h_{n+1}}{h_n},$$

$$\alpha_{2,2} = \int_0^1 \frac{x_n + sh_{n+1} - x_n}{x_{n-1} - x_n} ds = -\frac{1}{2} \frac{h_{n+1}}{h_n}.$$

For constant step size h these coefficients are $3/2$ and $-1/2$, respectively, so that equation (4) becomes:

$$y_{n+1} = y_n + h[\tfrac{3}{2}f_n - \tfrac{1}{2}f_{n-1}].$$

The second order method is not self-starting since there is no obvious way to get y_1. This illustrates a difficulty of all Adams methods of order $k > 1$, that is, how to obtain the starting values $y_1, y_2, \ldots, y_{k-1}$. Later we shall examine a way to resolve this difficulty by varying the step size and order.

The approximation, y_{n+1}, in (3) has two sources of error. One is due to approximating $y'(x)$ by an interpolating polynomial; this is known as the local truncation error. The other is due to the errors already present in the memorized values $y_n, y_{n-1}, \ldots$. We shall study the local truncation error now and use it in the next chapter to prove results about convergence.

The fundamental identity for the solution of equations (1) and (2) is

$$y(x_{n+1}) = y(x_n) + \int_{x_n}^{x_{n+1}} y'(t) dt.$$

Let $P_{k,n}(x)$ be the polynomial of degree $k - 1$ interpolating to $y'(x)$ at the mesh points:

$$P_{k,n}(x_{n+1-i}) = y'(x_{n+1-i}) = f(x_{n+1-i}, y(x_{n+1-i})) \qquad i = 1, \ldots, k.$$

The local truncation error τ_{n+1}^p is defined by

$$y(x_{n+1}) = y(x_n) + \int_{x_n}^{x_{n+1}} P_{k,n}(t) dt + \tau_{n+1}^p$$

$$= y(x_n) + h_{n+1} \sum_{i=1}^k \alpha_{k,i} f(x_{n+1-i}, y(x_{n+1-i})) + \tau_{n+1}^p.$$

Clearly

$$\tau^p_{n+1} = \int_{x_n}^{x_{n+1}} y'(t) - P_{k,n}(t) \, dt$$

is just the error due to interpolation. If $y(x) \in C^{k+1}[x_{n+1-k}, x_{n+1}]$, then Theorem 2 of Chapter 2 provides a useful expression for this error,

$$y'(x) - P_{k,n}(x) = E(x) \prod_{i=1}^{k} (x - x_{n+1-i}),$$

where $E(x)$ is a continuous function of x. Thus

$$\tau^p_{n+1} = \int_{x_n}^{x_{n+1}} E(t) \prod_{i=1}^{k} (t - x_{n+1-i}) \, dt.$$

Since the product is of one sign on the interval of integration, an integral mean value theorem says there is a point ζ in the interval such that

$$\tau^p_{n+1} = E(\zeta) \int_{x_n}^{x_{n+1}} \prod_{i=1}^{k} (t - x_{n+1-i}) \, dt.$$

A more meaningful representation for τ^p_{n+1} follows from the expression for $E(x)$ developed in Chapter 2:

$$\tau^p_{n+1} = \frac{y^{(k+1)}(\zeta)}{k!} \int_{x_n}^{x_{n+1}} \prod_{i=1}^{k} (t - x_{n+1-i}) \, dt.$$

If the step size is a constant, h, we can introduce the variable $s = (t - x_n)/h$ to simplify this expression further to

$$\tau^p_{n+1} = y^{(k+1)}(\zeta) h^{k+1} \frac{1}{k!} \int_{0}^{1} \prod_{i=1}^{k} (s + i - 1) \, ds$$

$$= \gamma_k h^{k+1} y^{(k+1)}(\zeta).$$

In general if H is the maximum step size considered, we have

$$\tau^p_{n+1} = \frac{y^{(k+1)}(\zeta)}{k!} H^{k+1} \int_{x_n}^{x_{n+1}} \prod_{i=1}^{k} \left(\frac{t - x_{n+1-i}}{H} \right) \frac{dt}{H}$$

so

$$|\tau^p_{n+1}| \le \frac{|y^{(k+1)}(\zeta)|}{k!} H^{k+1} \int_{x_n}^{x_{n+1}} \prod_{i=1}^{k} i \frac{dt}{H} \le |y^{(k+1)}(\zeta)| H^{k+1}.$$

Thus τ_{n+1}^p is $O(H^{k+1})$ for all n if $y(x)$ is sufficiently smooth.

The basic idea behind estimating the error of interpolation is to compare the result of an interpolating polynomial of one degree to that of a higher degree. This is valid when the mesh spacing is "small." The same idea works here, since

$$\tau_{n+1}^p = \int_{x_n}^{x_{n+1}} y'(t) - \mathcal{P}_{k,n}(t)\, dt$$

$$= \int_{x_n}^{x_{n+1}} \mathcal{P}_{k+1,n}(t) - \mathcal{P}_{k,n}(t)\, dt + \int_{x_n}^{x_{n+1}} y'(t) - \mathcal{P}_{k+1,n}(t)\, dt$$

and

$$\int_{x_n}^{x_{n+1}} y'(t) - \mathcal{P}_{k+1,n}(t)\, dt = O(H^{k+2})$$

can be ignored in estimating τ_{n+1}^p which is $O(H^{k+1})$. This estimate is not practical since we do not know $\mathcal{P}_{k,n}(t)$ and $\mathcal{P}_{k+1,n}(t)$. In the past a heuristic approach of ignoring the errors in the memorized values has been used. For if we presume $y_{n+1-i} = y(x_{n+1-i})$, then $\mathcal{P}_{k,n}(t) = P_{k,n}(t)$ and $\mathcal{P}_{k+1,n}(t) = P_{k+1,n}(t)$. With this assumption,

$$\tau_{n+1}^p = \int_{x_n}^{x_{n+1}} P_{k+1,n}(t) - P_{k,n}(t)\, dt + O(H^{k+2})$$

can be readily estimated when using the divided difference form of the Adams-Bashforth formula; τ_{n+1}^p is simply estimated to be the difference between y_{n+1} obtained using the formula of order k and that obtained from the formula of order $k + 1$. When the step size is constant we see from (5) that this estimate is

$$\tau_{n+1}^p \doteq h\gamma_k \nabla^k f_n.$$

A particularly fortunate aspect of this estimate is that if we keep enough differences, we can trivially adjust the order used so as to make the estimated local truncation error less than some tolerance. In Chapter 6 error estimation is carefully examined and the conclusion is that the approach is valid even when we account for the errors in the memorized values.

This analysis of the truncation error assumes that $y(x) \in C^{k+1}$ in order to get a tidy expression for the error of interpolation $y'(x) - \mathcal{P}_{k,n}(x)$. If $y(x)$ has fewer than $k + 1$ continuous derivatives, the order of τ_{n+1}^p is reduced accordingly. Nearly all real problems have solutions which are smooth, except at isolated points, and we assume this to be the situation throughout the text.

These expressions for the local truncation error indicate that high order formulas approximate the solution better than do low order ones for small step sizes. One might expect this since high order formulas use more interpolation points in defining $P_{k,n}(x)$. In the case of variable step sizes it is clear that reducing the step size will tend to reduce the error. The factor $y^{(k+1)}(\xi)$ should not be overlooked. If $y(x)$ varies slowly over the interval of interpolation, then the highest order derivatives are small and the truncation error is small. If $y(x)$ changes rapidly in this interval, the contrary is true. And so it is the slowly varying solutions which are relatively easy to obtain numerically.

It seems plausible that a better approximation to $y(x_{n+1})$ could be obtained if we regard the Adams-Bashforth value y_{n+1} of (3) as a tentative, "predicted" value and incorporate it into an interpolating polynomial. To distinguish the predicted value of (3) we shall write it henceforth as p_{n+1}. The Adams-Moulton formula of order k at x_n uses a polynomial $P^*_{k,n}(x)$ that also interpolates to k derivative values,

$$P^*_{k,n}(x_{n+1-j}) = f_{n+1-j} \qquad j = 1, \ldots, k-1,$$
$$P^*_{k,n}(x_{n+1}) = f(x_{n+1}, p_{n+1}).$$

The approximate solution, y_{n+1}, is given by

$$y_{n+1} = y_n + \int_{x_n}^{x_{n+1}} P^*_{k,n}(t)\,dt. \tag{7}$$

The Lagrangian form of (7) is derived exactly as in the Adams-Bashforth case; one finds that

$$y_{n+1} = y_n + h_{n+1} \sum_{i=1}^{k-1} \alpha^*_{k,i} f_{n+1-i} + h_{n+1} \alpha^*_{k,0} f(x_{n+1}, p_{n+1}) \tag{8}$$

where

$$\alpha^*_{k,i} = \int_0^1 l^*_i(x_n + sh_{n+1})\,ds$$

and

$$l^*_i(x) = \prod_{\substack{j=0 \\ j \neq i}}^{k-1} \frac{x - x_{n+1-j}}{x_{n+1-i} - x_{n+1-j}}.$$

The classical backward difference form of (7) is derived in the same way as for equation (5):

$$y_{n+1} = y_n + h \sum_{i=1}^{k} \gamma_{i-1}^* \nabla^{i-1} f_{n+1}^p \tag{9}$$

where

$$\gamma_0^* = 1,$$

$$\gamma_i^* = \frac{1}{i!} \int_0^1 (s-1)(s) \cdots (s+i-2) \, ds \qquad i \ge 1, \tag{10}$$

and

$$\nabla^0 f_{n+1}^p = f_{n+1}^p = f(x_{n+1}, p_{n+1}),$$

$$\nabla^1 f_{n+1}^p = \nabla f_{n+1}^p = f_{n+1}^p - f_n,$$

$$\vdots \qquad \qquad \vdots$$

$$\nabla^i f_{n+1}^p = \nabla^{i-1} f_{n+1}^p - \nabla^{i-1} f_n.$$

The case of variable step size is treated in Chapter 5.

EXERCISE 2

By considering the case of $f(x, y) \equiv 1$, show that for $k = 1, 2, \ldots$

$$\sum_{i=1}^{k} \alpha_{k,i} = 1 \quad \text{and} \quad \sum_{i=0}^{k-1} \alpha_{k,i}^* = 1.$$

EXERCISE 3

Prove that $\gamma_i^* = \gamma_i - \gamma_{i-1}$ for $i \ge 1$.

The local truncation error τ_{n+1} is defined by

$$y(x_{n+1}) = y(x_n) + h_{n+1} \sum_{i=0}^{k-1} \alpha_{k,i}^* f(x_{n+1-i}, y(x_{n+1-i})) + \tau_{n+1}.$$

In general

$$\tau_{n+1} = \frac{y^{(k+1)}(\eta)}{k!} \int_{x_n}^{x_{n+1}} \prod_{i=0}^{k-1} (t - x_{n+1-i}) \, dt,$$

and when the step size is a constant h,

$$\tau_{n+1} = \gamma_k^* h^{k+1} y^{(k+1)}(\eta).$$

In any case τ_{n+1} is $O(H^{k+1})$ for all n if $y(x)$ is sufficiently smooth. If one proceeds heuristically by assuming no error in the memorized or predicted values, then τ_{n+1} can be estimated as the difference of the results y_{n+1} of the formulas of order $k+1$ and of order k. For constant step size this estimate is

$$\tau_{n+1} \doteq h\gamma_k^* \nabla^k f_{n+1}^p.$$

Estimates of this kind will be justified by proving in Chapter 6 that the effect of errors in the memorized and predicted values can be ignored.

Conventionally the work involved in doing an integration on a computer is measured by the number of evaluations of $f(x, y)$ required. For complicated functions the remaining computation, the overhead, is relatively small so this practice is a convenient way to measure work which is also machine independent. For simple functions the measure is misleading, but then the whole computation is usually inexpensive and the comparison unimportant. The predictor-corrector procedure costs two evaluations at each step. Why not use the Adams-Bashforth method which costs only one function evaluation at each step? The predictor-corrector approach is more accurate (though not of a higher order) and so much better with respect to the propagation of error (see Chapter 8) that it can use steps more than twice as large. Also, the error estimates are more reliable, which leads to a more effective selection of the step size.

The kind of predictor-corrector procedure presented in this text is called a PECE method, an acronym derived from a description of how the computation is done: we Predict p_{n+1}, Evaluate f_{n+1}^p, Correct to get y_{n+1}, and Evaluate f_{n+1} to complete the step. We can use predictor and corrector formulas of different orders but as it turns out, in a PECE mode there is no advantage in using a corrector formula that is more than one order higher than the predictor. We shall consider two possibilities. One is to take both predictor and corrector of order k and the other is to take the predictor of order k and the corrector of order $k+1$. This latter possibility is not inconvenient, since the predictor of order k uses the values y_n and f_{n+1-j} for $j = 1, \ldots, k$ and the corrector of order $k+1$ uses the same values along with p_{n+1}. Thus the quantities to be retained are the same for either possibility. Several practical advantages which accrue by using the higher order corrector are discussed in succeeding chapters. Because of these advantages we are mainly interested in this possibility. When we refer to the PECE formulas of order k we shall always mean the Adams-Bashforth predictor of order k and the Adams-Moulton corrector of order $k+1$. This terminology may seem unnatural in the next chapter when the theory of convergence is discussed, but it is quite natural in Chapters 6 and 7 when the practical estimation and control of local error are discussed.

Let us look at a couple of simple cases. For $k = 1$ the Adams-Moulton corrector has $P^*_{1,n}(x)$ a constant, namely $f(x_{n+1}, p_{n+1})$, so

$$y_{n+1} = y_n + h_{n+1} f^p_{n+1}.$$

This is called the backward Euler method. If we combine it with Euler's method as predictor, the PECE formulas are

$$p_{n+1} = y_n + h_{n+1} f_n,$$
$$y_{n+1} = y_n + h_{n+1} f(x_{n+1}, p_{n+1}), \tag{11}$$
$$f_{n+1} = f(x_{n+1}, y_{n+1}).$$

Similarly when $k = 2$ the Adams-Moulton formula is

$$y_{n+1} = y_n + \frac{h_{n+1}}{2} [f(x_{n+1}, p_{n+1}) + f_n],$$

which is called the trapezoidal rule. If we use Euler's method ($k = 1$) as predictor and the trapezoidal rule ($k = 2$) as corrector, the PECE formulas are

$$p_{n+1} = y_n + h_{n+1} f_n,$$
$$y_{n+1} = y_n + \frac{h_{n+1}}{2} [f(x_{n+1}, p_{n+1}) + f_n], \tag{12}$$
$$f_{n+1} = f(x_{n+1}, y_{n+1}).$$

In view of the close relationship between the predictor and corrector formulas, we might hope that the computation of the corrected value would be facilitated by the previous computation of the predicted value. Using divided differences, this proves to be true. The general case will be treated in Chapter 5. At this point we shall mention only the case of constant step size h. If the predictor and corrector are both of order k, then

$$y_{n+1} = p_{n+1} + h\gamma_{k-1} \nabla^k f^p_{n+1}. \tag{13}$$

If the predictor is of order k and the corrector of order $k + 1$, then

$$y_{n+1} = p_{n+1} + h\gamma_k \nabla^k f^p_{n+1}. \tag{14}$$

These identities are easily proved from the formulas (5) and (9) and the result of Exercise 3.

Prove (13) and (14).

We have not yet defined the truncation error of the predictor-corrector process. After doing this in Chapter 6 and studying its estimation, we shall find that it is estimated by $h\gamma_k^*\nabla^k f_{n+1}^p$ when the step size is constant. In general the estimate is a multiple of the term to be added to p_{n+1} to get y_{n+1}. And so, the forms (13) and (14) are very efficient because, in any event, both terms in them must be computed.

The code provided starts an integration by using (12), which requires the use of only the differential equation and the initial condition. Since a PECE method requires two function evaluations per step, regardless of the order, and since we expect higher order formulas to be more efficient, we shall now return to the problem of obtaining starting values for higher order formulas. On completion of the first step we shall have stored y_1, f_1, and f_0, which is exactly the information that is needed for a second order predictor and third order corrector to take the second step. With y_2, f_2, f_1, and f_0 we could use a third order predictor and fourth order corrector, and so on. In this way we could increase the order as the necessary data is computed. The lower order formulas are generally less accurate, which must be compensated for by taking smaller steps. The ability to vary the step size and order which serves to start the code is used to good effect after the start by changing the step size and order to get the desired accuracy as cheaply as possible. However, any gain due to changing the step size must be balanced against the cost of computing the coefficients for the formulas. We shall study an efficient way to do this in Chapter 5, but we expect to use a constant step size most of the time.

4

CONVERGENCE
AND STABILITY—
SMALL STEP SIZES

The Adams formulas of Chapter 3 produce a value y_{n+1} which we expect will approximate the exact solution $y(x_{n+1})$. It is not at all clear that it does so since y_{n+1} depends on y_n and $f_n, f_{n-1}, \ldots$, all of which are normally in error. These errors propagate as the computation proceeds and can conceivably be amplified so that y_{n+1} is not even close to the exact solution. Many methods that are very similar to the Adams methods cannot be used for just this reason. We prove here that for the Adams methods in PECE form there is no computational difficulty because of the preceding errors and that the approximations converge to the exact solution as the step sizes tend to zero.

In this chapter the reader will begin to see the difference between theory and practice. The classical convergence and stability results are for constant step size and a fixed order formula. They are obtained by limit arguments that are based on letting the step size tend to zero, and so their implications are valid when the step size is "small." In practice both the step size and the order are manipulated to obtain the desired local accuracy as efficiently as possible. This is done during the course of a single integration and requires a viewpoint somewhat different from the classical one. Recent advances in theory and a suitable view of the classical results largely justify the practical approach. The attitude of this text is that codes like the one provided can, for the most part, be described as working with a constant step size and order (especially step size) with occasional changes. The analysis of this and succeeding chapters supports such a code adequately, though not completely. The theory must be complemented by carefully designed numerical experiments, such as those described in Chapter 11, and experience with real problems, especially those simple enough to permit some analysis, such as those described in Chapter 12.

Because we are usually interested in constant step size and order, we shall first develop the classical results. The proofs are quite simple because we deal only with Adams methods in PECE form. After discussing the principal reason for using a variable step formulation we modify the proofs a little so that they will apply to this case. We then discuss variation of the order.

And so we begin with an approximation y_{n+1} determined by the Adams-Bashforth-Moulton PECE combination with predictor of order k and corrector of order $k + 1$. A fixed step size h is used throughout the computation and starting values $y_0, y_1, \ldots, y_{k-1}$ have been obtained somehow. The relevant Adams formulas are

$$p_{n+1} = y_n + h \sum_{j=1}^{k} \alpha_{k,j} f(x_{n+1-j}, y_{n+1-j}), \tag{1a}$$

$$y_{n+1} = y_n + h \sum_{j=1}^{k} \alpha^*_{k+1,j} f(x_{n+1-j}, y_{n+1-j}) + h\alpha^*_{k+1,0} f(x_{n+1}, p_{n+1}). \tag{1b}$$

The local truncation errors τ^p_{n+1}, τ_{n+1} for the Adams-Bashforth and Adams-Moulton formulas, respectively, were defined in Chapter 3 by the relations

$$y(x_{n+1}) = y(x_n) + h \sum_{j=1}^{k} \alpha_{k,j} f(x_{n+1-j}, y(x_{n+1-j})) + \tau^p_{n+1}, \tag{2a}$$

$$y(x_{n+1}) = y(x_n) + h \sum_{j=0}^{k} \alpha^*_{k+1,j} f(x_{n+1-j}, y(x_{n+1-j})) + \tau_{n+1}. \tag{2b}$$

The global error at x_i is the discrepancy between the exact solution at x_i and the computed approximation:

$$e_i = y(x_i) - y_i.$$

By subtracting (1b) from (2b) and using a mean value theorem we see how the global error at x_{n+1} depends on the global errors at preceding mesh points:

$$e_{n+1} = e_n + h \sum_{j=1}^{k} \alpha^*_{k+1,j} g_{n+1-j} e_{n+1-j} + h\alpha^*_{k+1,0} g_{n+1}\left[y(x_{n+1}) - p_{n+1}\right]$$
$$+ \tau_{n+1},$$

where $g_{n+1-j} = f_y(x_{n+1-j}, \xi_{n+1-j})$ for some ξ_{n+1-j} between y_{n+1-j} and $y(x_{n+1-j})$ for $j > 0$; and between p_{n+1} and $y(x_{n+1})$ for $j = 0$. We need not be more precise about g_{n+1-j} because we shall bound it by the Lipschitz constant L; recall that

$$|f_y(x, y)| \leq L \qquad \text{for } a \leq x \leq b \text{ and all } y.$$

The relation involves the error in the predicted value, $y(x_{n+1}) - p_{n+1}$, which (in the same way) we find to be

$$e^p_{n+1} = y(x_{n+1}) - p_{n+1} = e_n + h \sum_{j=1}^{k} \alpha_{k,j} g_{n+1-j} e_{n+1-j} + \tau^p_{n+1}.$$

Eliminating this term in the expression for e_{n+1} we arrive at

$$e_{n+1} = e_n + h \sum_{j=1}^{k} \alpha^*_{k+1,j} g_{n+1-j} e_{n+1-j} \qquad (3)$$
$$+ h\alpha^*_{k+1,0} g_{n+1}\left[e_n + h \sum_{j=1}^{k} \alpha_{k,j} g_{n+1-j} e_{n+1-j}\right] + \delta_n.$$

We shall call the quantity

$$\delta_n = h\alpha^*_{k+1,0} g_{n+1} \tau^p_{n+1} + \tau_{n+1}$$

the local truncation error of the predictor-corrector combination. If we use $\delta = \max_m |\delta_m|$ and the Lipschitz constant, then

$$|e_{n+1}| \leq |e_n| + hL \sum_{j=1}^{k} |\alpha^*_{k+1,j}| |e_{n+1-j}| + hL |\alpha^*_{k+1,0}| |e_n|$$
$$+ h^2 L^2 |\alpha^*_{k+1,0}| \sum_{j=1}^{k} |\alpha_{k,j}| |e_{n+1-j}| + \delta.$$

The bound, in this form, offers little insight as to the nature and magnitude of the error. We need to replace it with a simpler expression which is more easily interpreted. To do this let us introduce a sequence of numbers $\{E_i\}$ satisfying

$$|e_j| \leq E_i \qquad \text{for } j = 0, 1, \ldots, i \text{ and all } i,$$

and assume we have found the first terms $E_0, E_1, \ldots, E_n$ for $n \geq k-1$. Then

$$|e_{n+1}| \leq E_n + hL \sum_{j=1}^{k} |\alpha_{k+1,j}^*| E_n + hL |\alpha_{k+1,0}^*| E_n$$

$$+ h^2 L^2 |\alpha_{k+1,0}^*| \sum_{j=1}^{k} |\alpha_{k,j}| E_n + \delta.$$

So if we define E_{n+1} by

$$E_{n+1} = \mathcal{L} E_n + \delta$$

where

$$\mathcal{L} = 1 + hL\alpha^* + h^2 L^2 |\alpha_{k+1,0}^*| \alpha,$$

$$\alpha^* = \sum_{j=0}^{k} |\alpha_{k+1,j}^*|, \quad \text{and} \quad \alpha = \sum_{j=1}^{k} |\alpha_{k,j}|,$$

then it can be seen that

$$|e_{n+1}| \leq E_{n+1}.$$

Furthermore we see

$$|e_j| \leq E_n \leq E_{n+1} \quad \text{for } j = 0, 1, \ldots, n$$

by the assumption on E_n and the fact that the recipe for E_{n+1} makes it larger than E_n. To begin the sequence, all we need to do is define

$$E_0 = \max \{|e_0|, |e_1|, \ldots, |e_{k-1}|\},$$

$$E_i = \mathcal{L} E_{i-1} + \delta \quad i = 1, 2, \ldots, k-1.$$

The remaining task is to explicitly determine the E_i and to state the result in a simple form. We use a couple of lemmas for this purpose.

LEMMA   1   Let $\{E_i\}$ be a sequence defined by

$$E_0 \text{ given,}$$

$$E_k = \mathcal{L} E_{k-1} + \delta \quad k = 1, 2, \ldots,$$

where $\mathcal{L}, \delta$ are positive constants and $\mathcal{L} \neq 1$. Then

$$E_k = \mathcal{L}^k E_0 + \frac{\mathcal{L}^k - 1}{\mathcal{L} - 1} \delta \quad k = 1, 2, \ldots.$$

Proof.   The proof is by induction on k. The assertion is trivially true for

$k = 1$. Assume it is true for $k = n$. Then

$$E_{n+1} = \mathscr{L}E_n + \delta$$

$$= \mathscr{L}\left[\mathscr{L}^n E_0 + \frac{\mathscr{L}^n - 1}{\mathscr{L} - 1}\delta\right] + \delta$$

$$= \mathscr{L}^{n+1}E_0 + \frac{\mathscr{L}^{n+1} - 1}{\mathscr{L} - 1}\delta,$$

which is the case $k = n + 1$. By induction the assertion is true for all k.

LEMMA   2   For $x \geq 0$,
   a) $1 + x \leq \exp(x)$
   b) $(1 + x)^n \leq \exp(nx)$
where equality holds only if $x = 0$.

Proof. A Taylor series expansion of $\exp(x)$ about $x = 0$ yields

$$\exp(x) = 1 + x + \frac{x^2}{2}\exp(\xi) \qquad \xi \in (0, x)$$

and the lemma follows immediately.

Using Lemmas 1 and 2 we can now state:

THEOREM   1   Let the predictor-corrector scheme (1) be applied to solve $y' = f(x, y)$, $y(a) = A$ in $[a, b]$ with starting values y_i satisfying

$$|y(x_i) - y_i| \leq E_0 \qquad i = 0, 1, \ldots, k - 1.$$

Let $f_y(x, y)$ be continuous and bounded in magnitude by the Lipschitz constant L. Then for any $x_n \in [a, b]$

$$|y(x_n) - y_n| \leq \left[E_0 + \frac{\delta}{hL(\alpha^* + hL|\alpha^*_{k+1,0}|\alpha)}\right] \times$$
$$\exp[(x_n - a)L(\alpha^* + hL|\alpha^*_{k+1,0}|\alpha)] \qquad (4)$$

where

$$\delta = \max_n |h\alpha^*_{k+1,0}f_y \tau^p_{n+1} + \tau_{n+1}|,$$

$$\alpha = \sum_{j=1}^k |\alpha_{k,j}|, \qquad \alpha^* = \sum_{j=0}^k |\alpha^*_{k+1,j}|.$$

Proof: We have argued that

$$|y(x_n) - y_n| \le E_n.$$

Lemma 1 applies to the E_n used in this argument, so

$$E_n = \mathcal{L}^{\,n} E_0 + \frac{\mathcal{L}^{\,n} - 1}{\mathcal{L} - 1} \delta.$$

If we take

$$x = hL\alpha^* + h^2 L^2 |\alpha^*_{k+1,0}| \alpha$$

in Lemma 2 and use the fact that $nh = x_n - a$, we see

$$\mathcal{L}^{\,n} \le \exp\left[(x_n - a)L(\alpha^* + hL|\alpha^*_{k+1,0}|\alpha)\right].$$

The conclusion of the theorem is now evident.

This theorem implies the convergence of the PECE pair. Of the several kinds of convergence, we discuss uniform convergence which means simply that the worst error in $[a, b]$,

$$\max_{a \le x_n \le b} |y(x_n) - y_n|,$$

tends to zero. Inequality (4) gives a bound on the worst error, but to interpret it we need to investigate the terms a little further. In Chapter 3 we found that if $y(x) \in C^{k+2}[a, b]$, then τ^p_{n+1} is $O(h^{k+1})$ and τ_{n+1} is $O(h^{k+2})$. This implies δ is $O(h^{k+2})$, which leads to the following result:

COROLLARY   1.1   In addition to the assumptions of Theorem 1 suppose that $y(x) \in C^{k+2}[a, b]$ and that the starting values satisfy

$$|y(x_i) - y_i| \le c_1 h^{k+1} \qquad i = 0, 1, \ldots, k - 1$$

for some constant c_1 and all $0 < h \le h_0$. Then there is a constant c_2 such that for any $x_n \in [a, b]$

$$|y(x_n) - y_n| \le c_2 h^{k+1},$$

that is, the scheme is uniformly convergent of order $k + 1$.

EXERCISE 1

Investigate the effect of using predictors and correctors of different orders. Suppose the predictor is of order k, the corrector is of order m, the solution $y(x)$ is as smooth as necessary, and the starting values are as accurate as necessary. Prove the PECE scheme is uniformly convergent of order $p = \min(m, k + 1)$. (It is for this reason that only the cases $m = k$ and $m = k + 1$ are considered in this text.)

Since the PECE procedure of order k is convergent of order $k + 1$, the terminology seems a bit unnatural here. However, it is natural from other points of view, one of those being that k memorized values are used. The only difference apparent so far due to using the higher order corrector is that not doing so requires an additional starting value and an additional memorized value at each step to get the same accuracy. This is a relatively minor point, but it favors the higher order corrector.

The classical convergence result of Corollary 1.1 applies only to approximation at the mesh points. In Chapter 3 we saw how to approximate $y(x)$ at any point in $[a, b]$. We shall now show that such an approximation is as accurate everywhere as it is at the mesh points. This has the extremely important practical consequence that codes can select mesh points for the sake of efficiency and still obtain accurate solution values anywhere by interpolation.

The typical situation for interpolating the solution is to integrate to the first mesh point x_{n+1} beyond the output point x at which we want the solution, i.e. $x_n < x \le x_{n+1}$. The value y_{n+1} is obtained using $P^*_{k+1,n}$ and then f_{n+1} is evaluated. A natural approximation to y' is $P_{k+1,n+1}$, which differs from $P^*_{k+1,n}$ only in that it interpolates to the accepted value f_{n+1}, rather than to f^p_{n+1}. We define an approximation $y_I(x)$ to $y(x)$ on $(x_n, x_{n+1}]$ by

$$y_I(x) = y_{n+1} + \int_{x_{n+1}}^{x} P_{k+1,n+1}(t)\, dt.$$

First we examine the approximation of $y'(x)$ by $P_{k+1,n+1}(x)$. Let $\mathcal{P}_{k+1,n+1}(x)$ be the polynomial of degree k defined by the conditions

$$\mathcal{P}_{k+1,n+1}(x_{n+2-i}) = y'(x_{n+2-i}) \qquad i = 1, 2, \ldots, k + 1.$$

Then

$$\left| y'(x) - P_{k+1,n+1}(x) \right| \le \left| y'(x) - \mathcal{P}_{k+1,n+1}(x) \right| + \left| \mathcal{P}_{k+1,n+1}(x) - P_{k+1,n+1}(x) \right|.$$

Theorem 2 of Chapter 2 says that

$$\left| y'(x) - \mathcal{P}_{k+1,n+1}(x) \right| = \left| \frac{y^{(k+2)}(\xi)}{(k+1)!} \prod_{i=1}^{k+1} (x - x_{n+2-i}) \right| \le h^{k+1} \left\| y^{(k+2)} \right\|$$

where $\left\| y^{(k+2)} \right\| = \max\limits_{a \le x \le b} \left| y^{(k+2)}(x) \right|$. The term

$$\mathcal{P}_{k+1,n+1}(x) - P_{k+1,n+1}(x) = \sum_{i=1}^{k+1} l_i(x) \left[f(x_{n+2-i}, y(x_{n+2-i})) \right.$$
$$\left. - f(x_{n+2-i}, y_{n+2-i}) \right]$$

so, from Corollary 1.1,

$$\left| \mathcal{P}_{k+1,n+1}(x) - P_{k+1,n+1}(x) \right| \le L c_2 h^{k+1} \sum_{i=1}^{k+1} \left| l_i(x) \right|.$$

Using the variable $s = (x - x_{n+1})/h$ it is easy to see from the definition of $l_i(x)$ that

$$\left| l_i(x) \right| = \prod_{\substack{j=1 \\ j \ne i}}^{k+1} \left| \frac{s+j-1}{j-i} \right| \le \prod_{\substack{j=1 \\ j \ne i}}^{k+1} \left| \frac{j}{j-i} \right|.$$

Thus there are constants c_3 and c_4 such that

$$\left| \mathcal{P}_{k+1,n+1}(x) - P_{k+1,n+1}(x) \right| \le c_3 h^{k+1}$$

and

$$\left| y'(x) - P_{k+1,n+1}(x) \right| \le h^{k+1} \left\| y^{(k+2)} \right\| + c_3 h^{k+1} = c_4 h^{k+1}.$$

Using this result and the identity

$$y(x) = y(x_{n+1}) + \int_{x_{n+1}}^{x} y'(t)\, dt$$

we then see that for a suitable constant c_5

$$\left| y(x) - y_I(x) \right| \le \left| y(x_{n+1}) - y_{n+1} \right| + \left| \int_{x_{n+1}}^{x} y'(t) - P_{k+1,n+1}(t)\, dt \right|$$

$$\le c_2 h^{k+1} + h c_4 h^{k+1} \le c_5 h^{k+1}.$$

COROLLARY   1.2   With the assumptions of Theorem 1 and Corollary 1.1, the numerical approximation $y_I(x)$ and its derivative are uniformly convergent of order h^{k+1} on $[a, b]$ to $y(x)$ and its derivative, respectively.

EXERCISE 2

When $k = 1$ verify that for $x_n \le x \le x_{n+1}$,

$$y_I'(x) = f_{n+1} + (x - x_{n+1})\left(\frac{f_{n+1} - f_n}{h_{n+1}}\right),$$

$$y_I(x) = y_{n+1} + (x - x_{n+1})f_{n+1} + \frac{(x - x_{n+1})^2}{2}\left(\frac{f_{n+1} - f_n}{h_{n+1}}\right).$$

The quantity E_0 in Theorem 1 represents the starting errors which, according to the error bound, apparently persist throughout the computation. A more delicate analysis shows this is the case and it is easily demonstrated by experiment. (See Exercise 5.) And so one must start accurately, a difficulty that is examined later. There is a satisfactory practical solution but it does not fit neatly into the classical limit analysis and it is better analyzed from a viewpoint developed in Chapter 6. It is usually realistic to suppose that the starting errors contribute little to the global error. For this reason we shall make some theoretical applications of Corollary 1.1 assuming the starting errors are $O(h^{k+2})$. The global error is then still of order $k + 1$ but the starting errors prove to be negligible compared to the truncation errors.

The quantity δ in Theorem 1 represents the error in approximating the differential equation, and in principle can be made arbitrarily small. In practice the Adams formulas (1) cannot be evaluated exactly. Rather, there are roundoff errors, r_n^p and r_n, that must be incorporated into the expression δ_n at each step. In doing so we can no longer say δ goes to zero with h because roundoff errors do not decrease with h, indeed δ may increase. Ordinarily accuracies and step sizes are used with magnitudes chosen in such a way that roundoff errors are small compared to truncation errors, and can thus be ignored both in theory and practice. But as more and more accuracy is requested, it is eventually found that the roundoff errors dominate the truncation errors. It then becomes necessary to incorporate r_n^p and r_n into δ in the theory to account for the fact that in practice the numerical results typically become worse instead of better. In Chapter 9 we examine some effects of roundoff in solving differential equations and also consider some simple devices for their control. In computing practice we must presume that roundoff will cause no serious problems, while keeping in mind that this is certainly not true when we approach accuracies near those limits determined by the computer being used.

The convergence results can be misleading when applied to practical computation, quite aside from considerations of roundoff. It appears from these results that the highest order formula feasible ought to be used. This conclusion is true in a limit analysis as h tends to 0 but for each single computation it is not, for then the unknown constants in the analysis play a decisive role. After starting, the error results mainly from the maximum truncation error δ. In Chapter 6 we discuss the estimation of local errors at each step for several orders while allowing a variation of step size. Granted that this is feasible and economic, one suspects from Theorem 1 that the step size and order can be adapted to the solution at each step so as to reduce δ_n, hence δ and $|y(x_n) - y_n|$. Or, put the other way around, one should be able to use steps as large as possible and still get a specified error, as turns out to be the case. The remaining development of convergence results in the text is to accommodate changes in step size and order.

The use of methods similar to the Adams methods can result in the computations "blowing up" on some problems as the step size is reduced. This is called "instability," which unfortunately is a term used to describe many different phenomena both in the theory of differential equations and in the theory of their numerical solution. The Adams methods do not exhibit this behavior. In the simplest form we ask that the numerical values y_n be uniformly bounded for all sufficiently small h, that is, that they do not "blow up" when an integration is repeated with a smaller h. With the assumptions of Corollary 1.1 this boundedness is immediate. For in this case there are constants c and h_0 such that for all $h \leq h_0$,

$$|y(x_n) - y_n| \leq ch^{k+1}$$

because of convergence. But then

$$|y_n| \leq ch_0^{k+1} + \max_{a \leq x \leq b} |y(x)|.$$

A much more precise statement is that small changes in the starting values lead to small changes in the numerical solution:

THEOREM   2   Suppose that $f(x, y)$ satisfies a Lipschitz condition for $a \leq x \leq b$ and all y. Let $\{y_n\}$ be the numerical solution obtained by the formulas of (1) from starting values $y_0, y_1, \ldots, y_{k-1}$ and $\{u_n\}$ that obtained from starting values $u_0, u_1, \ldots, u_{k-1}$. There are constants B and h_0 such that if $0 < h \leq h_0$, then

$$\max_{x_n \in [a, b]} |y_n - u_n| \leq B \max_{0 \leq j \leq k-1} |y_j - u_j|.$$

This result is easily proved by an argument much like that used to prove Theorem 1.

EXERCISE 3

Prove Theorem 2.

A very simple and cheap way to change the step size is by interpolation. For a method of order k let f_{n-i} be the computed derivatives at $x_n - ih$ for $i = 0, 1, \ldots, k - 1$. If the step size in going to x_{n+1} is to be changed by a factor r so that $x_{n+1} = x_n + rh$, a polynomial interpolating the f_{n-i} is found and evaluated at $x_n - i \cdot rh$ to produce equally spaced data points $\tilde{f}_{n-i}$. The code then proceeds as usual, applying the constant step formula to the $\tilde{f}_{n-i}$. This step adjustment procedure is very convenient but it can be extremely unstable. The instability is not fully understood, but it depends on the magnitudes of changes, whether or not the step size is being decreased, and on the order of the formula. It is troublesome only when rather high order formulas are being used (e.g., 7 and up) and thus the instability was not clearly recognized until quite recently, when codes taking advantage of high order methods became generally available. The most powerful and elegant way to avoid this very serious practical difficulty is to use formulas incorporating variable step sizes, as is done in this text.

Experience suggests that the ratio of successive step sizes will need to be bounded in order to get stability. The proof of Theorem 1 uses constants like

$$\alpha = \sum_{j=1}^{k} |\alpha_{k,j}|.$$

In the variable step size case these will be replaced by bounds. From the example

$$\sum_{j=1}^{2} |\alpha_{2,j}| = 1 + \frac{1}{2} \frac{h_{n+1}}{h_n} + \frac{1}{2} \frac{h_{n+1}}{h_n},$$

it appears that some sort of uniform bound on the ratio of successive step sizes will be needed to prove a result analogous to Theorem 1.

The stability of a completely variable step size PECE formulation of the Adams methods is not difficult to establish once we see a suitable way to limit changes in step size. The arguments for fixed step size h consider a limit process in which an integration is repeated for successively smaller step sizes. Analogous arguments for variable step size assume that the integration is repeated with successively smaller maximum step sizes $H = \max h_j$. We must assume that the mesh points $\{x_0, x_1, \ldots, x_N\}$ actually span the interval $[a, b]$, so we require

$$b - a = \sum_{j=1}^{N} h_j. \tag{5a}$$

In the case of constant step size we used the bound $nh = x_n - a \leq b - a$ and in the general case we need an analogous assumption:

there is a constant D such that for each integration $NH \leq D$.   (5b)

We have suggested that the ratio of successive step sizes should be bounded. Specifically, we suppose that

there are positive, finite constants μ, η such that, for all integrations,

$$\mu \leq h_{j-1}/h_j \leq \eta \tag{5c}$$

for all j except, possibly, those for which h_j is taken at order one.

No bound on the ratio of step sizes is needed at order one because the formulas have no memorized values. This exception is of considerable practical importance.

These assumptions are enough to carry out the analysis and appear to be entirely adequate for describing real computation. They do not amount to assuming that the step sizes tend to a uniform value, for it can be seen that the ratio of maximum to minimum step size can be unbounded (*without* using the exception at order one). The codes provided satisfy assumption (5c) with $\mu = \frac{1}{2}$ and $\eta = 8$. The usual way of gaining confidence in the results of an actual computation is to repeat it with a smaller error tolerance, or with a succession of smaller tolerances. We shall see that this must normally result in a succession of smaller maximum step sizes.

Our first task is to show that the $\alpha_{k,i}$ and $\alpha_{k+1,i}^*$ are uniformly bounded with these rules about the step size. Recall that

$$\alpha_{k,i} = \int_0^1 l_i(x_n + sh_{n+1})\,ds, \qquad l_i(x) = \prod_{\substack{j=1 \\ j \neq i}}^{k} \left(\frac{x - x_{n+1-j}}{x_{n+1-i} - x_{n+1-j}} \right),$$

hence

$$|\alpha_{k,i}| \leq \int_0^1 \prod_{\substack{j=1 \\ j \neq i}}^{k} \left| \frac{x_n + sh_{n+1} - x_{n+1-j}}{x_{n+1-i} - x_{n+1-j}} \right| ds.$$

To show that each factor is bounded, divide the numerator and denominator by h_{n+1}. Then assumption (5c) and the fact that $0 \leq s \leq 1$ imply that for $j > 1$

$$\left| \frac{x_n + sh_{n+1} - x_{n+1-j}}{h_{n+1}} \right| \leq \left| \frac{x_{n+1} - x_{n+1-j}}{h_{n+1}} \right| = \left| \frac{h_{n+1} + \cdots + h_{n+2-j}}{h_{n+1}} \right|$$

$$\leq 1 + \eta + \eta^2 + \cdots + \eta^{j-1}$$

since for $m = 0, 1, \ldots$

$$\frac{h_{n-m}}{h_{n+1}} = \frac{h_{n-m}}{h_{n+1-m}} \cdot \frac{h_{n+1-m}}{h_{n+2-m}} \cdots \frac{h_n}{h_{n+1}} \leq \eta^{m+1}.$$

The same bound can also be seen to hold for $j = 1$. The denominator is bounded in much the same way, the specific bound depending on the relative sizes of i and j. For $j > i$

$$\left| \frac{x_{n+1-i} - x_{n+1-j}}{h_{n+1}} \right| = \left| \frac{h_{n+1-i} + h_{n-i} + \cdots + h_{n+2-j}}{h_{n+1}} \right|$$

which is bounded below by

$$\mu^i + \mu^{i+1} + \cdots + \mu^{j-1}.$$

Using the identity

$$1 + \sigma + \sigma^2 + \cdots + \sigma^{i-1} = \frac{\sigma^i - 1}{\sigma - 1}, \qquad \sigma \neq 1,$$

we find that each factor with $j > i$ is bounded above by

$$\frac{\dfrac{\eta^j - 1}{\eta - 1}}{\dfrac{\mu^j - \mu^i}{\mu - 1}}.$$

There are similar bounds for $j < i$ and consequently $|l_i(x_n + sh_{n+1})|$ is bounded independently of the mesh spacing, as is $|\alpha_{k,i}|$. In exactly the same way we find that the $\alpha^*_{k+1,i}$ are also uniformly bounded.

Only a few details in the proof of Theorem 1 need to be changed for the proof to apply to variable step sizes. Equation (3) remains true, with the additional notation of h_{n+1}:

$$e_{n+1} = e_n + h_{n+1} \sum_{j=1}^{k} \alpha^*_{k+1,j} g_{n+1-j} e_{n+1-j}$$

$$+ h_{n+1} \alpha^*_{k+1,0} g_{n+1} \left[e_n + h_{n+1} \sum_{j=1}^{k} \alpha_{k,j} g_{n+1-j} e_{n+1-j} \right] + \delta_n. \tag{6}$$

Now we take α and α^* to be constants such that

$$\alpha^* \geq \sum_{j=0}^{k} |\alpha^*_{k+1,j}|, \qquad \alpha \geq \sum_{j=1}^{k} |\alpha_{k,j}| \tag{7}$$

for all mesh spacings considered. Then using a maximum step size H and defining a constant C such that

$$|\alpha^*_{k+1,0}| \leq C, \tag{8}$$

we find

$$|e_{n+1}| \leq E_n(1 + HL\alpha^* + H^2 L^2 C\alpha) + \delta, \tag{9}$$

exactly as was found previously. In applying the lemmas we have expressions like nH which are now bounded by $nH \leq NH \leq D$. In this way we are led to the following theorem:

THEOREM 3   Let the predictor-corrector scheme (1) be applied to solve $y' = f(x, y)$, $y(a) = A$ in $[a, b]$ with starting values y_i satisfying

$$|y(x_i) - y_i| \leq E_0 \qquad i = 0, 1, \ldots, k - 1.$$

Let $f_y(x, y)$ be continuous and bounded in magnitude by the Lipschitz constant L. Consider a sequence of integrations with assumptions (5a, b, c) holding. Then for any $x_n \in [a, b]$

$$|y(x_n) - y_n| \leq \left[E_0 + \frac{\delta}{HL(\alpha^* + HLC\alpha)} \right] \exp[DL(\alpha^* + HLC\alpha)]. \tag{10}$$

We have discussed the truncation error of the predictor and corrector formulas for variable steps in Chapter 3. The proofs of the following corollaries are left to the reader.

COROLLARY 3.1   In addition to the assumptions of Theorem 3, suppose that $y(x) \in C^{k+2}[a, b]$ and that the starting values satisfy

$$|y(x_i) - y_i| \leq c_1 H^{k+1} \qquad i = 0, 1, \ldots, k-1$$

for some constant c_1 and all $0 < H \leq H_0$. Then there is a constant c_2 such that for any $x_n \in [a, b]$,

$$|y(x_n) - y_n| \leq c_2 H^{k+1}.$$

COROLLARY 3.2   With the assumptions of Theorem 3 and Corollary 3.1, the numerical approximation $y_I(x)$ and its derivative are uniformly convergent of order H^{k+1} on $[a, b]$ to $y(x)$ and its derivative, respectively.

As we have indicated, we also intend to vary the order to keep δ small in the error bound (10). By redefining the constants used previously to account for this the analog of Theorem 3 is easily proved. Suppose we allow those orders k with $1 \leq k \leq K$; the code provided has $K = 12$. Let α^*, α, and C be constants such that equations (7) and (8) are true for all orders in this range. Equation (6) is also true as long as there are enough data points already computed to apply the formula of order k. Then the inequality (9) is true at every step. In this way we arrive at the following theorem:

THEOREM 4   Let the predictor-corrector scheme (1) be applied to solve $y' = f(x, y)$, $y(a) = A$ in $[a, b]$ with $1 \leq k \leq K$. Any order k may be used at any step, provided enough points have been computed to apply the formula. Suppose a formula of order k_0 is used to start and starting values y_i are given which satisfy

$$|y(x_i) - y_i| \leq E_0 \qquad i = 0, 1, \ldots, k_0 - 1.$$

Let $f_y(x, y)$ be continuous and bounded in magnitude by the Lipschitz constant L. Consider a sequence of integrations with assumptions (5a, b, c) holding. Then for any $x_n \in [a, b]$, the inequality (10) holds.

Just as in the case of fixed step size and order we have stability in the sense that Theorem 4 implies that the numerical solutions are uniformly bounded. The more precise statement of Theorem 2 also has its analog, which is proved in the same way.

THEOREM   5   Suppose that $f(x, y)$ satisfies a Lipschitz condition for $a \le x \le b$ and all y. Let the predictor-corrector scheme (1) be applied with any order $k \le K$ at any step, provided that enough points have been computed to apply the formula. Suppose a formula of order k_0 is used to start, and that $\{y_n\}$ is the numerical solution obtained from starting values $y_0, y_1, \ldots, y_{k_0-1}$, and that $\{u_n\}$ is obtained from the starting values $u_0, u_1, \ldots, u_{k_0-1}$. There are constants B and H_0 such that if $0 < H \le H_0$, then

$$\max_{x_n \in [a, b]} |y_n - u_n| \le B \max_{0 \le j \le k_0 - 1} |y_j - u_j|.$$

EXERCISE 4

The equations $y' = f(x, y)$ are said to satisfy a conservation law if there is a constant vector v such that for any solution $y(x)$ of the equations, $v^T y(x) \equiv c$ for some constant c determined by the initial conditions. Such a law is inherent in the form of the differential equations. It arises by integrating the identity $v^T f(x, y) \equiv 0$, which is assumed to hold for all y. A given system of equations may satisfy several such laws.

The problem

$$y' = My = \begin{pmatrix} -1 & & & & \\ 1 & -2 & & \mathbf{0} & \\ & 2 & -3 & & \\ & & & \ddots & \\ & \mathbf{0} & & 8 & -9 \\ & & & & 9 & 0 \end{pmatrix} y, \quad y(0) = \begin{pmatrix} 1 \\ 0 \\ 0 \\ \vdots \\ 0 \\ 0 \end{pmatrix}$$

describing a radioactive decay chain has a conservation law with $v^T = (1, 1, \ldots, 1)$ since $v^T y' = v^T M y = \mathbf{0}^T y \equiv 0$. Thus the sum of the species is a constant which from the initial conditions must be 1.

Prove that if the PECE methods of order k with $1 \le k \le K$ are used (any order may be used as long as there is enough data to apply the formulas) to solve a system with a conservation law of this type, then in exact arithmetic the numerical solution also satisfies the conservation law. For example, the numerical solution of the radioactive decay chain problem has $v^T y_n \equiv 1$ for all n.

We have established the convergence of the variable order, variable step Adams methods but the error bound we derived gives no indication of the actual errors. We state here and prove below an estimate of the error for constant step size and order when the starting errors are $O(h^{k+2})$:

$$y(x_n) = y_n + h^{k+1}\phi(x_n) + O(h^{k+2})$$

where the function $\phi(x)$ is defined as the solution of

$$\phi' = G(x)\phi + \gamma^*_{k+1}y^{(k+2)}(x) + \gamma_k \alpha^*_{k+1,0}G(x)y^{(k+1)}(x),$$

$$\phi(a) = 0$$

and $G(x) = f_y(x, y(x))$. The bound is an increasing function of x because it must be expected that from all the problems in the class considered, at least one leads to an actual growth in the error. The estimate shows that the behavior of the error is quite complex. Many people, on first encounter, expect the error to grow as the computation proceeds: but this need not be so, it can decrease as well. Let us look at a numerical example which illustrates both this fact and the estimate of the error stated above.

Let us solve

$$y' = -y,$$

$$y(0) = 1$$

with the case $k = 1$, that is, by using formulas (12) of Chapter 3, the combination of Euler's method and the trapezoidal rule. Here, $y(x) = \exp(-x)$ and $f_y \equiv -1$ so

$$\phi' = -\phi + \left(-\frac{1}{12}\right)(-\exp(-x)) + \left(\frac{1}{2}\right)\left(\frac{1}{2}\right)(-1)(\exp(-x)).$$

(The coefficients γ_1, γ^*_1 may be obtained from their definitions.) The solution of

$$\phi' = -\phi - \frac{1}{6}\exp(-x),$$

$$\phi(0) = 0$$

is easily verified to be

$$\phi(x) = \frac{-1}{6}x \exp(-x).$$

We have asserted that the error

$$y(x_n) - y_n = \frac{-h^2}{6} x_n \exp(-x_n) + O(h^3).$$

The function $|\phi(x)|$ increases until $x = 1$, then decreases, so we see that for small h the absolute error behaves in the same way. Because we know $y(x)$ we also see that

$$\left| \frac{y(x_n) - y_n}{y(x_n)} \right| = \frac{h^2}{6} x_n + O(h^3).$$

that is, the relative error keeps on increasing. To verify this formula numerically we compute

$$\Delta(h) = \frac{y_n - y(x_n)}{h^2} \qquad \text{at } x_n = 1$$

for a sequence of h tending to zero in such a way that $x_n = 1$ for some n, namely $h = 10^{-1}, 10^{-2}, \dots$. According to our error estimate we should have

$$\Delta(h) \rightarrow \frac{1}{6} \exp(-1) = 0.0613$$

as shown in the following table.

m	$\Delta(10^{-m}) \times 10^2$
1	6.62
2	6.18
3	6.14
4	6.13
$\vdots$	$\vdots$
∞	6.13

EXERCISE 5

Repeat this computation with $y_0 = 1 + h$. Because $y(x_0) - y_0$ is $O(h)$ you will find $y(x_n) - y_n$ is $O(h)$ at $x_n = 1$, showing that the starting error persists. To see this numerically compute both $\Delta(h)$ and

$$\delta(h) = \frac{y_n - y(x_n)}{h}$$

and examine them as $h \rightarrow 0$.

The following table shows the errors for fixed $h = 10^{-2}$ for various x to demonstrate that the errors can also decrease, as well as increase in the course of an integration.

x_n	$(y_n - y(x_n)) \times 10^6$
0.0	0
0.2	2.75
0.4	4.50
0.6	5.53
0.8	6.04
1.0	6.18
1.2	6.07
1.4	5.80
1.6	5.42
1.8	5.00
2.0	4.55

These expressions for the error can illustrate the gain to be obtained by proper choice of step size and order. To be specific, if we use a predictor of order k and corrector of order $k + 1$ to solve

$$y' = y, \qquad y(0) = 1,$$

then

$$y(x_n) - y_n = \Gamma_k x_n \exp(x_n) h^{k+1} + O(h^{k+2})$$

where

$$\Gamma_k = \gamma_{k+1}^* + \gamma_k \alpha_{k+1,0}^*.$$

An interesting situation arises when we compute with the error at x relative to $x \exp(x)$ for then

$$\frac{y(x_n) - y_n}{x_n \exp(x_n)} \doteq \Gamma_k h^{k+1}.$$

The largest step size that can be used, while still obtaining an error of ε, is determined by

$$\varepsilon \doteq |\Gamma_k| h^{k+1},$$

so

$$h \doteq \left(\frac{\varepsilon}{|\Gamma_k|}\right)^{\frac{1}{k+1}}.$$

Generally the best step size for obtaining a given error ε at x depends on x, but we see that with this way of measuring error for this problem, it does not. And so the h estimated here is optimal for all x, meaning that the problem ought to be integrated with a fixed step size. The assumption of fixed step size for convenience in the analysis is, in this case, also dictated by efficiency. For the same reason a fixed order ought to be used; the optimal order is that k leading to the largest h for a given ε. To see the kind of effect that varying the order has, we tabulate, as shown in the following table, the h associated with various orders and tolerances:

	ε	
k	10^{-4}	10^{-10}
1	.02	.00002
2	.09	.00009
3	.17	.0054
4	.25	.016
5	.32	.032
6	.38	.053
7	.43	.077
8	.48	.10
9	.52	.13
10	.55	.16
11	.58	.18
12	.61	.21

A good choice of order means that the problem can be solved with a much larger step size, hence much more efficiently, than with a poor choice. This table also illustrates the fact that as more accuracy is requested, higher order formulas become more appropriate. For this particular problem a rather high order is desirable, even for modest accuracy, but from a practical point of view the "best" order is poorly determined.

ASYMPTOTIC DEVELOPMENT OF ERROR

In this optional section we give some of the details of a proof that if $f(x, y)$ is sufficiently differentiable and if the starting values are sufficiently accurate, then the error of the PECE procedure of order k can be written as

$$e_n = y(x_n) - y_n = h^{k+1}\phi(x_n) + O(h^{k+2})$$

when the step size is a constant h. Here $\phi(x)$ is the solution of

$$\phi'(x) = G(x)\phi(x) + \alpha_{k+1,0}^* G(x)\gamma_k y^{(k+1)}(x) + \gamma_{k+1}^* y^{(k+2)}(x),$$
$$\phi(a) = 0$$

where $y(x)$ is the solution of $y' = f(x, y)$, $y(a)$ is given, and we let

$$G(x) = f_y(x, y(x)).$$

In our sketch of the proof of this statement we assume that $f(x, y)$ is smooth. The analytical complications caused by assuming limited differentiability cannot be justified by the practical use we make of the result; careful treatment of this kind of asymptotic development can be found in references [2, 12, and 26].

The proof involves a reappraisal of the expressions developed in the first part of the chapter—a reappraisal that takes into account the result of Corollary 1.1 that e_n is $O(h^{k+1})$. The first observation is that

$$e_{n+1}^p = e_n + h \sum_{j=1}^{k} \alpha_{k,j} g_{n+1-j} e_{n+1-j} + \tau_{n+1}^p$$

$$= e_n + h^{k+1} \gamma_k y^{(k+1)}(\eta) + O(h^{k+2})$$

for some η between x_{n+1-k} and x_{n+1}. Using this in the corresponding expression for e_{n+1} we find

$$e_{n+1} = e_n + h \sum_{j=0}^{k} \alpha_{k+1,j}^* g_{n+1-j} e_{n+1-j}$$

$$+ h^{k+2} \alpha_{k+1,0}^* g_{n+1} \gamma_k y^{(k+1)}(\eta) + h^{k+2} \gamma_{k+1}^* y^{(k+2)}(\xi) + O(h^{k+3})$$

where ξ is also between x_{n+1-k} and x_{n+1}. If we apply the corrector formula to $\phi(x)$ and use the equation it satisfies, we see that

$$\phi(x_{n+1}) = \phi(x_n) + h \sum_{j=0}^{k} \alpha_{k+1,j}^* [G(x_{n+1-j})\phi(x_{n+1-j})$$

$$+ \alpha_{k+1,0}^* G(x_{n+1-j}) \gamma_k y^{(k+1)}(x_{n+1-j}) + \gamma_{k+1}^* y^{(k+2)}(x_{n+1-j})] + O(h^{k+2}).$$

The result to be proved is that $\varepsilon_n = e_n/h^{k+1} - \phi(x_n)$ is $O(h)$. To clarify the relation between e_{n+1} and $\phi(x_{n+1})$ we note that

$$g_{n+1-j} = f_y(x_{n+1-j}, \xi_{n+1-j}) = G(x_{n+1-j}) + O(h^{k+1})$$

because ξ_{n+1-j} is between $y(x_{n+1-j})$ and either y_{n+1-j} or p_{n+1}. In a similar vein

$$G(x_{n+1-j}) y^{(k+1)}(x_{n+1-j}) = G(x_{n+1}) y^{(k+1)}(x_{n+1}) + O(h)$$

for $j = 1, \ldots, k$. In this manner we find that

$$e_{n+1} = e_n + h \sum_{j=0}^{k} \alpha_{k+1,j}^* G(x_{n+1-j}) e_{n+1-j}$$

$$+ h^{k+2} \alpha_{k+1,0}^* G(x_{n+1}) \gamma_k y^{(k+1)}(x_{n+1}) + h^{k+2} \gamma_{k+1}^* y^{(k+2)}(x_{n+1}) + O(h^{k+3})$$

and

$$\phi(x_{n+1}) = \phi(x_n) + h \sum_{j=0}^{k} \alpha_{k+1,j}^* G(x_{n+1-j}) \phi(x_{n+1-j})$$

$$+ h \alpha_{k+1,0}^* G(x_{n+1}) \gamma_k y^{(k+1)}(x_{n+1}) + h \gamma_{k+1}^* y^{(k+2)}(x_{n+1}) + O(h^2).$$

(To get the expression for $\phi(x_{n+1})$ we also use the identities of Exercise 2 of Chapter 3.)

From these expressions it is an easy step to

$$\varepsilon_{n+1} = \varepsilon_n + h \sum_{j=0}^{k} \alpha_{k+1,j}^* G(x_{n+1-j}) \varepsilon_{n+1-j} + O(h^2).$$

We presume that the starting values y_n are correct to order h^{k+2}. Since $\phi(a) = 0$ we have $\phi(x_n) = nh\phi'(a) + O(h^2)$ for $n = 0, 1, \ldots, k-1$ hence

$$\varepsilon_n = \frac{y(x_n) - y_n}{h^{k+1}} - \phi(x_n) = O(h) \qquad \text{for } n = 0, 1, \ldots, k-1.$$

An argument along the lines of Theorem 1 now shows that ε_n is $O(h)$ for all $x_n \in [a, b]$, which is the result we wanted.

5

EFFICIENT IMPLEMENTATION OF THE ADAMS METHODS

The principal cost of solving a differential equation lies in the repeated evaluation of the equation. The Adams methods, when carefully used, are more efficient in this respect than any other method being used today to solve general differential equations. To achieve this efficiency it is necessary to vary the step size and the order that are used. Accordingly it is necessary to estimate the errors that are, or would be, incurred for various step sizes and orders so as to make these decisions. Potent codes also attempt to detect abnormal situations such as discontinuities or certain kinds of instabilities and to deal with them in a reasonable way. The cost of all this computation is referred to as the overhead. As one might expect, variable step, variable order Adams codes in general, and especially ones designed to provide protection and convenience to the user, have a considerable overhead. The implementation—that is, the way the interpolating polynomials are represented and how the computation is organized—can affect this cost greatly. These decisions have an equally great effect on the simplicity and clarity of the code. In this chapter a very efficient implementation of the Adams methods which is also very well adapted to the tasks of the code is developed.

As observed in earlier chapters, the formulas for advancing a step do not require computation of their coefficients when the step size is constant, and so the overhead is then low. Most steps are taken in groups with a constant step size and constant order; this is particularly true of the step size. F. T. Krogh [18] has given an implementation which takes advantage of this fact to reduce the overhead a great deal in comparison to always computing the necessary coefficients. It is based on the divided difference form of the interpolating polynomial. There are several reasons for thinking that this might be a good approach. Our study of this form in Chapter 2 suggests it is especially convenient for advancing a step and for changing the order of interpolation. Since the error estimates developed in the next chapter amount to comparing the results of formulas of different orders, we expect the divided difference form to lend itself to error estimation. Krogh's implementation reduces to the classical backward difference formulation of the Adams methods when the step sizes are constant. This classical formulation has proved itself to be an economical one with good roundoff properties. Since most steps will be taken with a constant step size, it seems likely that this implementation of variable step formulas will share these properties of the classical formulation. On the other hand, interpolating the solution to get its value at output points turns out to be less convenient than for other implementations. On balance Krogh's implementation appears superior to any other for variable order Adams codes that permit complete variability of the step size while anticipating that much of the computation will be done with a constant step size.

The kth order Adams-Bashforth predictor at x_n is defined by the expression

$$y_n + \int_{x_n}^{x} P_{k,n}(t)\,dt \tag{1}$$

where $P_{k,n}(x)$ satisfies the interpolation conditions

$$P_{k,n}(x_{n+1-j}) = f_{n+1-j} \qquad j = 1,\ldots,k.$$

Expression (1) is used to predict both the solution and its derivative at x_{n+1}:

$$p_{n+1} = y_n + \int_{x_n}^{x_{n+1}} P_{k,n}(t)\,dt, \tag{2a}$$

$$p'_{n+1} = P_{k,n}(x_{n+1}). \tag{2b}$$

Later in this chapter we shall see why it is useful to introduce the predicted derivative p'_{n+1}.

Obviously we must determine an easy way to integrate the interpolating polynomial. As will be seen this will involve a good deal of notation and manipulation, but the final result is quite economical computationally. We start with the polynomial in divided difference form

$$P_{k,n}(x) = f[x_n] + (x - x_n)f[x_n, x_{n-1}] + (x - x_n)(x - x_{n-1})f[x_n, x_{n-1}, x_{n-2}]$$

$$+ \cdots + (x - x_n)(x - x_{n-1})\cdots(x - x_{n-k+2})f[x_n, x_{n-1}, \ldots, x_{n-k+1}]. \quad (3)$$

It is entirely possible to take the mesh points $x_n, x_{n-1}, \ldots$ and the divided differences as the fundamental quantities to be used in the code. But because we intend to vary the step size, it seems a little more natural to take the step sizes, $h_i = x_i - x_{i-1}$, and sums of the step sizes, $\psi_i(n + 1) = h_{n+1} + h_n + \cdots + h_{n+2-i} = x_{n+1} - x_{n+1-i}$, as fundamental quantities. The divided differences will be modified so that they reduce to backward differences when the step sizes are constant. Using these modified divided differences, $\phi_i(n)$, leads to economical, simple formulas. For integrations like that in (2a) we again find it convenient to introduce a normalized variable s by $x = x_n + sh_{n+1}$ so that s runs from 0 to 1 as x runs from x_n to x_{n+1}. Two other fundamental quantities, $\alpha_i(n + 1)$ and $\beta_i(n + 1)$, will arise naturally in the development of the algorithm. The basic definitions of the quantities used by the code are:

$$h_i = x_i - x_{i-1},$$

$$s = \frac{x - x_n}{h_{n+1}},$$

$$\psi_i(n + 1) = h_{n+1} + h_n + \cdots + h_{n+2-i} \qquad i \geq 1,$$

$$\alpha_i(n + 1) = \frac{h_{n+1}}{\psi_i(n + 1)} \qquad i \geq 1,$$

$$\beta_1(n + 1) = 1,$$

$$\beta_i(n + 1) = \frac{\psi_1(n + 1)\psi_2(n + 1)\cdots\psi_{i-1}(n + 1)}{\psi_1(n)\psi_2(n)\cdots\psi_{i-1}(n)} \qquad i > 1,$$

$$\phi_1(n) = f[x_n] = f_n,$$

$$\phi_i(n) = \psi_1(n)\psi_2(n)\cdots\psi_{i-1}(n)f[x_n, x_{n-1}, \ldots, x_{n-i+1}] \qquad i > 1.$$

$$(4)$$

It is worth noting that if the step size is a constant h, then $\psi_i(n + 1) = ih$, $\alpha_i(n + 1) = 1/i$, $\beta_i(n + 1) = 1$, and $\phi_i(n) = \nabla^{i-1} f_n$ for each i. In Chapter 2 we discussed efficient ways to compute divided differences the application of which will result in efficient ways to compute the $\phi_i(n)$. The other expressions in the definitions (4) above also lend themselves to computation since we can change from the values for n to those for $n + 1$ by computing for $i = 2, 3, \ldots,$

$$\psi_i(n + 1) = \psi_{i-1}(n) + h_{n+1},$$

$$\beta_i(n + 1) = \beta_{i-1}(n + 1) \cdot \frac{\psi_{i-1}(n + 1)}{\psi_{i-1}(n)},$$

after starting with the initial values

$$\psi_1(n + 1) = h_{n+1}, \qquad \beta_1(n + 1) = 1.$$

Note that we take the modified divided difference $\phi_i(n)$ to be associated with the $(i - 1)$th divided difference. In this respect the index is clumsy but it agrees with a natural FORTRAN implementation, which is more important.

Returning to the matter of integrating the interpolating polynomial, we find that in the notation of (4) a typical term of $P_{k,n}(x)$,

$$(x - x_n)(x - x_{n-1}) \cdots (x - x_{n-i+2}) f[x_n, x_{n-1}, \ldots, x_{n-i+1}], \tag{5}$$

may be written

$$(sh_{n+1}) \cdot (sh_{n+1} + h_n) \cdots (sh_{n+1} + h_n + \cdots + h_{n-i+3}) \cdot \frac{\phi_i(n)}{\psi_1(n)\psi_2(n) \cdots \psi_{i-1}(n)}$$

$$= \left(\frac{sh_{n+1}}{\psi_1(n + 1)}\right) \cdot \left(\frac{sh_{n+1} + h_n}{\psi_2(n + 1)}\right) \cdots \left(\frac{sh_{n+1} + h_n + \cdots + h_{n-i+3}}{\psi_{i-1}(n + 1)}\right) \cdot$$

$$\frac{\psi_1(n + 1)\psi_2(n + 1) \cdots \psi_{i-1}(n + 1)}{\psi_1(n)\psi_2(n) \cdots \psi_{i-1}(n)} \phi_i(n)$$

$$= \left(\frac{sh_{n+1}}{\psi_1(n + 1)}\right) \cdot \left(\frac{sh_{n+1} + \psi_1(n)}{\psi_2(n + 1)}\right) \cdots \left(\frac{sh_{n+1} + \psi_{i-2}(n)}{\psi_{i-1}(n + 1)}\right) \cdot \beta_i(n + 1)\phi_i(n).$$

To condense this a bit further we introduce for analytical purposes the quantities

$$c_{i,n}(s) = \begin{cases} 1 & i = 1, \\ \dfrac{sh_{n+1}}{\psi_1(n+1)} = s & i = 2, \quad (6) \\ \left(\dfrac{sh_{n+1}}{\psi_1(n+1)}\right) \cdot \left(\dfrac{sh_{n+1} + \psi_1(n)}{\psi_2(n+1)}\right) \cdots \left(\dfrac{sh_{n+1} + \psi_{i-2}(n)}{\psi_{i-1}(n+1)}\right) & i \geq 3. \end{cases}$$

If we also let

$$\phi_i^*(n) = \beta_i(n+1)\phi_i(n),$$

the typical term (5) may be written

$$c_{i,n}(s)\phi_i^*(n)$$

hence

$$P_{k,n}(x) = \sum_{i=1}^{k} c_{i,n}(s)\phi_i^*(n). \qquad (7)$$

To approximate the derivative of the solution by (2b) we take $s = 1$ in equation (7). From definitions (4) and (6)

$$c_{i,n}(1) = 1 \qquad i = 1, 2, \ldots, k,$$

so that we find the convenient representation

$$P_{k,n}(x_{n+1}) = p'_{n+1} = \sum_{i=1}^{k} \phi_i^*(n).$$

To approximate the solution at x_{n+1}, we substitute equation (7) into equation (2a) and integrate to obtain

$$p_{n+1} = y_n + h_{n+1} \sum_{i=1}^{k} \phi_i^*(n) \int_0^1 c_{i,n}(s)\,ds. \qquad (8)$$

This formula is the generalization to variable step sizes of the classical backward difference formulation

$$p_{n+1} = y_n + h \sum_{i=1}^{k} \gamma_{i-1} \nabla^{i-1} f_n,$$

and reduces it to when the step size is a constant h.

The integration of $c_{i,n}(s)$ requires some further computation in the case of variable step sizes and we use a circuitous approach to develop suitable formulas. Substituting the definitions of (4) into equation (6) we may write

$$c_{i,n}(s) = \begin{cases} 1 & i = 1, \\ \alpha_1(n+1)s = s & i = 2, \\ \left[\alpha_{i-1}(n+1)s + \dfrac{\psi_{i-2}(n)}{\psi_{i-1}(n+1)} \right] c_{i-1,n}(s) & i \geq 3. \end{cases}$$

Let us fix n and $i \geq 3$. From the expression for $c_{i,n}(s)$ and an integration by parts we find

$$\int_0^s c_{i,n}(s_0) \, ds_0 = \int_0^s \left[\alpha_{i-1}(n+1)s_0 + \frac{\psi_{i-2}(n)}{\psi_{i-1}(n+1)} \right] c_{i-1,n}(s_0) \, ds_0$$

$$= \left[\alpha_{i-1}(n+1)s + \frac{\psi_{i-2}(n)}{\psi_{i-1}(n+1)} \right] \int_0^s c_{i-1,n}(s_0) \, ds_0 \qquad (9)$$

$$- \int_0^s \alpha_{i-1}(n+1) \int_0^{s_1} c_{i-1,n}(s_0) \, ds_0 \, ds_1.$$

This seems to be of no value at all since we have introduced a double integral, but on closer examination it will be found that the subscript $i-1$ which has arisen will eventually prove this approach to be useful.

Let us use the notation for the q-fold integral of $c_{i,n}(s)$,

$$c_{i,n}^{(-q)}(s) = \int_0^s \int_0^{s_{q-1}} \cdots \int_0^{s_1} c_{i,n}(s_0)\, ds_0\, ds_1 \ldots ds_{q-1}.$$

In this notation equation (9) states that

$$c_{i,n}^{(-1)}(s) = \left[\alpha_{i-1}(n+1)s + \frac{\psi_{i-2}(n)}{\psi_{i-1}(n+1)}\right] c_{i-1,n}^{(-1)}(s) - \alpha_{i-1}(n+1)c_{i-1,n}^{(-2)}(s).$$

Repeated integration by parts leads to the general relation

$$c_{i,n}^{(-q)}(s) = \left[\alpha_{i-1}(n+1)s + \frac{\psi_{i-2}(n)}{\psi_{i-1}(n+1)}\right] c_{i-1,n}^{(-q)}(s) - q\alpha_{i-1}(n+1)c_{i-1,n}^{(-q-1)}(s).$$

To see this, we first observe that it holds for $q = 1$. Assuming it to be true for $q = m$, we substitute the relation into the definition of $c_{i,n}^{(-m-1)}(s)$ and integrate the first term by parts to get:

$$c_{i,n}^{(-m-1)}(s) = \int_0^s c_{i,n}^{(-m)}(s_0)\, ds_0 = \int_0^s \left[\alpha_{i-1}(n+1)s_0 + \frac{\psi_{i-2}(n)}{\psi_{i-1}(n+1)}\right] c_{i-1,n}^{(-m)}(s_0)\, ds_0$$

$$- \int_0^s m\alpha_{i-1}(n+1)c_{i-1,n}^{(-m-1)}(s_0)\, ds_0$$

$$= \left[\alpha_{i-1}(n+1)s + \frac{\psi_{i-2}(n)}{\psi_{i-1}(n+1)}\right]\int_0^s c_{i-1,n}^{(-m)}(s_0)\, ds_0$$

$$- \int_0^s \alpha_{i-1}(n+1)\int_0^{s_1} c_{i-1,n}^{(-m)}(s_0)\, ds_0\, ds_1$$

$$- m\alpha_{i-1}(n+1)c_{i-1,n}^{(-m-2)}(s),$$

or

$$c_{i,n}^{(-m-1)}(s) = \left[\alpha_{i-1}(n+1)s + \frac{\psi_{i-2}(n)}{\psi_{i-1}(n+1)}\right] c_{i-1,n}^{(-m-1)}(s)$$

$$- (m+1)\alpha_{i-1}(n+1)c_{i-1,n}^{(-m-2)}(s).$$

By induction the general relation holds. From equation (8) we see that only the coefficients $c_{i,n}^{(-1)}(1)$ are needed. They can be calculated from $c_{i,n}^{(-q)}(1)$ but, for reasons of scaling, it is better to define the quantities

$$g_{i,q} = (q-1)!\, c_{i,n}^{(-q)}(1).$$

If we substitute the value $s = 1$ in the relation just developed, we see that

$$c_{i,n}^{(-q)}(1) = \left[\alpha_{i-1}(n+1) + \frac{\psi_{i-2}(n)}{\psi_{i-1}(n+1)} \right] c_{i-1,n}^{(-q)}(1) - q\alpha_{i-1}(n+1)c_{i-1,n}^{(-q-1)}(1).$$

But then

$$(q-1)!\, c_{i,n}^{(-q)}(1) = \left[\frac{h_{n+1} + \psi_{i-2}(n)}{\psi_{i-1}(n+1)} \right] (q-1)!\, c_{i-1,n}^{(-q)}(1)$$

$$- \alpha_{i-1}(n+1)q!\, c_{i-1,n}^{(-q-1)}(1),$$

or

$$g_{i,q} = g_{i-1,q} - \alpha_{i-1}(n+1)g_{i-1,q+1}.$$

After dealing with the special cases of $i = 1$ and 2 we find that

$$g_{i,q} = \begin{cases} \dfrac{1}{q} & i = 1, \\[3mm] \dfrac{1}{q(q+1)} & i = 2, \\[3mm] g_{i-1,q} - \alpha_{i-1}(n+1)g_{i-1,q+1} & i \geq 3, \end{cases} \tag{10}$$

and the formula

$$p_{n+1} = y_n + h_{n+1} \sum_{i=1}^{k} g_{i,1}\phi_i^*(n). \tag{11}$$

EXERCISE 3
Verify directly that $c_{1,n}^{(-q)}(s) = s^q/q!$ so that $g_{1,q} = 1/q$. Also verify that $c_{2,n}^{(-q)}(s) = s^{q+1}/(q+1)!$ so that $g_{2,q} = 1/[q(q+1)]$.

EXERCISE 4
Observe that $c_{i,n}(s) > 0$ for $s > 0$, hence $c_{i,n}^{(-q)}(1) > 0$ and $g_{i,q} > 0$. Then prove that $g_{i-1,q} > g_{i,q}$. The fact that $g_{k,1} > g_{k+1,1}$ is used in one of the codes provided.

To see how the coefficients $g_{i,1}$ that are needed might be generated, suppose that we are working at order $k = 4$ and using a constant step size so that $\alpha_i(n + 1) = 1/i$. The coefficients $g_{i,q}$ form a triangular table with the first two columns generated by known values and all succeeding columns by the recursion (10), in this case,

$$g_{i,q} = g_{i-1,q} - \frac{g_{i-1,q+1}}{(i-1)}.$$

This table of coefficients is

$$
\begin{array}{c}
 & & & i & & \\
 & & 1 & 2 & 3 & 4 \\
\hline
 & 1 & 1 & 1/2 & 5/12 & 3/8 \\
 & 2 & 1/2 & 1/6 & 1/8 & \\
q & 3 & 1/3 & 1/12 & & \\
 & 4 & 1/4 & & &
\end{array}
$$

The short dark lines indicate which pairs of elements q and $q + 1$ in column $i - 1$ are used to generate a new element q in column i.

It will develop that the quantity $g_{k+1,1}$ is needed for the error estimate and the correction, so the triangular array will be extended to generate it, too. Because $g_{1,1} = 1$ for any step size and because the coefficients $g_{2,q} = 1/[q(q + 1)]$ can be used to initialize the array, in practice, we start with $i = 2$. Thus in the preceding example we would actually generate the following table:

$$
\begin{array}{c}
 & & & i & & \\
 & & 2 & 3 & 4 & 5 \\
\hline
 & 1 & 1/2 & 5/12 & 3/8 & 251/720 \\
 & 2 & 1/6 & 1/8 & 19/180 & \\
q & 3 & 1/12 & 7/120 & & \\
 & 4 & 1/20 & & &
\end{array}
$$

From Exercise 1 it is obvious that $\gamma_{i-1} = g_{i,1}$ when the step sizes are constant. This furnishes a simple way to compute the γ_i and also the $\gamma_i^* = \gamma_i - \gamma_{i-1}$ once and for all. The error estimation procedures require these coefficients so they are set in a DATA statement in the code.

Although the derivation was a bit tricky, the final algorithm for generating the coefficients needed in equation (8) for the predicted value p_{n+1} is computationally very simple. Later when we discuss the actual programming we shall find that it is quite economical of storage and uses groups of steps taken at a constant step size to reduce the work.

We are now ready to "correct" p_{n+1} and prepare for the next step. The code uses a corrector one order higher than that of the predictor and accepts as an approximation to the solution and its derivative at x_{n+1}

$$y_{n+1} = y_n + \int_{x_n}^{x_{n+1}} P_{k+1,n}^*(t)\, dt,$$

$$f_{n+1} = f(x_{n+1}, y_{n+1})$$

where

$$P_{k+1,n}^*(x_{n+1-j}) = f_{n+1-j} \qquad j = 1,\dots,k,$$

$$P_{k+1,n}^*(x_{n+1}) = f_{n+1}^p = f(x_{n+1}, p_{n+1}).$$

The corrector polynomial interpolates to the same data as the predictor plus the additional value f_{n+1}^p. For the error estimation we also want to consider the case of the corrector of order k, $P_{k,n}^*(x)$, being used. The analysis is much the same and the reader is advised to work out this case to test his understanding.

Since $P_{k+1,n}^*(x)$ interpolates the same data as $P_{k,n}(x)$ plus one extra point, the basic intent of the divided difference form is to represent it as a small correction to $P_{k,n}(x)$. Reference to Chapter 2 shows that

$$P_{k+1,n}^*(x) = P_{k,n}(x) + (x - x_n)(x - x_{n-1})\cdots(x - x_{n-k+1}) \cdot$$

$$f^p[x_{n+1}, \dots, x_{n-k+1}]. \quad (12)$$

The superscript p on this divided difference is to remind us that $P_{k+1,n}^*(x)$ interpolates to f_{n+1}^p. (If we were to use f_{n+1} here, we would have $P_{k+1,n+1}(x)$ instead of $P_{k+1,n}^*(x)$.)

We shall use a superscript p on all the divided differences associated with $P_{k+1,n}^*(x)$.

On introducing the new notation we find that

$$P_{k+1,n}^*(x) = P_{k,n}(x) + c_{k+1,n}(s)\phi_{k+1}^p(n + 1).$$

Integrating this equation we obtain

$$y_{n+1} = y_n + h_{n+1} \int_0^1 P_{k,n}(x_n + sh_{n+1})\, ds + h_{n+1} \int_0^1 c_{k+1,n}(s)\phi_{k+1}^p(n + 1)\, ds.$$

Thus

$$y_{n+1} = p_{n+1} + h_{n+1}g_{k+1,1}\phi_{k+1}^p(n+1),$$
$$f_{n+1} = f(x_{n+1}, y_{n+1}).$$
(13)

This is a very convenient way to compute y_{n+1} since we can compute the coefficient $g_{k+1,1}$ along with the $g_{i,1}$ needed in the prediction process. An entirely similar argument shows that if we use a corrector of order k instead of $k+1$, the only change in the formula for y_{n+1} is to change $g_{k+1,1}$ to $g_{k,1}$. We shall write $y_{n+1}(k)$ to distinguish this value:

$$y_{n+1}(k) = p_{n+1} + h_{n+1}g_{k,1}\phi_{k+1}^p(n+1).$$

These two formulas generalize to variable step size the formulas developed in Chapter 3 for constant step size:

$$y_{n+1} = p_{n+1} + h\gamma_k \nabla^k f_{n+1}^p,$$
$$y_{n+1}(k) = p_{n+1} + h\gamma_{k-1}\nabla^k f_{n+1}^p.$$

All that remains is to see how to compute the $\phi_i^p(n+1)$ and the $\phi_i(n+1)$ to complete the step. For $i > 1$ the basic relation (6) of Chapter 2 between divided differences gives

$$\phi_{i+1}^p(n+1) = \psi_1(n+1)\cdots\psi_i(n+1)f^p[x_{n+1},\ldots,x_{n-i+1}]$$
$$= \psi_1(n+1)\cdots\psi_{i-1}(n+1)f^p[x_{n+1},\ldots,x_{n-i+2}]$$
$$- \frac{\psi_1(n+1)\cdots\psi_{i-1}(n+1)}{\psi_1(n)\cdots\psi_{i-1}(n)}\cdot\psi_1(n)\cdots\psi_{i-1}(n)f[x_n,\ldots,x_{n-i+1}]$$
$$= \phi_i^p(n+1) - \beta_i(n+1)\phi_i(n),$$

that is,

$$\phi_{i+1}^p(n+1) = \phi_i^p(n+1) - \phi_i^*(n).$$
(14)

Since $\phi_1^p(n+1) = f_{n+1}^p$ by definition, we can see that the divided difference representation is extremely convenient for advancing one step; just use equation (14) to compute the $\phi_{i+1}^p(n+1)$ in the order $i = 1,\ldots,k$. Exactly the same argument shows that

$$\phi_{i+1}(n+1) = \phi_i(n+1) - \phi_i^*(n),$$
(15)

and using $\phi_1(n + 1) = f_{n+1}$ we can compute the $\phi_{i+1}(n + 1)$ just as we do the $\phi_{i+1}^p(n + 1)$.

EXERCISE 5

Define $g_{0,1} = 0$ and then prove the relation

$$y_{n+1} = y_n + h_{n+1} \sum_{i=1}^{k+1} (g_{i,1} - g_{i-1,1})\phi_i^P(n + 1),$$

which generalizes to variable step size the classical formula

$$y_{n+1} = y_n + h \sum_{i=1}^{k+1} \gamma_{i-1}^* \nabla^{i-1} f_{n+1}^P.$$

This is an extremely simple and cheap way to advance a step but it does not make good use of storage. There are a lot of differences to be retained and we should like to organize the computations so that they are as economical of storage as possible. It would be ideal if we could form $\phi_i^*(n)$ and store it where $\phi_i(n)$ was (overwrite); then form $\phi_i^P(n + 1)$ and store it where $\phi_i^*(n)$ was, and lastly form $\phi_i(n + 1)$ and store it where $\phi_i^P(n + 1)$, hence $\phi_i(n)$, was. We cannot do this if we use (14) in the order $i = 1, 2, \ldots, k$ but we can, if we rewrite (14) as

$$\phi_i^P(n + 1) = \phi_{i+1}^p(n + 1) + \phi_i^*(n) \tag{16}$$

and use (16) in the order $i = k, k - 1, \ldots, 1$. To get started we shall need $\phi_{k+1}^p(n + 1)$, but from (14) it is easy to see that

$$\phi_{k+1}^P(n + 1) = f_{n+1}^P - p_{n+1}' = f_{n+1}^P - \sum_{i=1}^{k} \phi_i^*(n).$$

If we proceed in this way, how can we obtain the $\phi_i(n + 1)$? The relationships (14) and (15) are so similar that we should expect some simple relation directly connecting $\phi_i(n + 1)$ and $\phi_i^P(n + 1)$. Subtracting (14) from (15) we see that

$$\phi_{i+1}(n + 1) - \phi_{i+1}^p(n + 1) = \phi_i(n + 1) - \phi_i^p(n + 1) = \cdots$$

$$= \phi_1(n + 1) - \phi_1^p(n + 1).$$

Then since $\phi_1(n + 1) - \phi_1^p(n + 1) = f_{n+1} - \phi_1^p(n + 1)$, we have

$$\phi_i(n + 1) = \phi_1^p(n + 1) + (f_{n+1} - \phi_1^p(n + 1)), \tag{17}$$

for each i.

This is a feasible approach which is economical of storage, but we shall use a variation that facilitates the computation of $\phi_{k+1}^p(n + 1)$ and also takes into account the fact that we need only a few of the $\phi_i^p(n + 1)$ for the error estimation. Equation (17) expresses the close relation of $P_{k+1,n}^*(x)$ and $P_{k,n+1}(x)$; the variation uses the equally close relation to $P_{k,n}(x)$. There is a slight complication due to the introduction of some intermediate quantities, $\phi_i^e(n + 1)$, which are just divided differences for $P_{k,n}(x)$ based on the mesh points $x_{n+1}, x_n, \ldots$. Let us think of $P_{k,n}(x)$ as a polynomial of degree k determined by the values

$$P_{k,n}(x_{n+1-j}) = f_{n+1-j} \qquad j = 1, \ldots, k,$$

$$P_{k,n}(x_{n+1}) = p_{n+1}'$$

and see how to determine the modified divided differences, $\phi_i^e(n + 1)$, corresponding to it. The superscript e is to remind us that these differences use the value p_{n+1}' at x_{n+1} so that we are extending the differences $\phi_i(n)$, based on values at $x_n, x_{n-1}, \ldots$, to differences $\phi_i^e(n + 1)$ based on values at $x_{n+1}, x_n, \ldots$. Just as in proving (14) we find that

$$\phi_{i+1}^e(n + 1) = \phi_i^e(n + 1) - \phi_i^*(n),$$

or, in the form we use it, that

$$\phi_i^e(n + 1) = \phi_{i+1}^e(n + 1) + \phi_i^*(n). \tag{18}$$

Since $P_{k,n}(x)$ is actually a polynomial of degree $k - 1$, all divided differences of order greater than $k - 1$ are zero. Thus

$$\phi_{k+1}^e(n + 1) = \psi_1(n + 1)\cdots\psi_k(n + 1)f^e[x_{n+1}, \ldots, x_{n-k+1}] = 0.$$

With this result we can use (18) to generate the $\phi_i^e(n + 1)$ in the sequence $i = k, k - 1, \ldots, 1$ and overwrite them on the $\phi_i^*(n)$. The trick used to prove the relation (17) can be used in the same way to prove

$$\phi_i^p(n + 1) = \phi_i^e(n + 1) + (f_{n+1}^p - \phi_1^e(n + 1)), \tag{19}$$

$$\phi_i(n + 1) = \phi_i^e(n + 1) + (f_{n+1} - \phi_1^e(n + 1)). \tag{20}$$

Using (19) we generate those $\phi_i^p(n + 1)$ required for the error estimation. If the step is successful, we then use (20) to compute the $\phi_i(n + 1)$ and overwrite them on the $\phi_i^e(n + 1)$.

Let us now summarize all the computation required to advance one step. The code that is given in Chapter 10 uses a PECE Adams method which advances from x_n to x_{n+1} by

computing $g_{i,1}$, $\qquad i = 1, 2, \ldots, k + 1$;

P   Predicting $\phi_i^*(n) = \beta_i(n + 1)\phi_i(n)$, $\qquad i = 1, 2, \ldots, k$,

$$p_{n+1} = y_n + h_{n+1} \sum_{i=1}^{k} g_{i,1}\phi_i^*(n),$$

$$\phi_{k+1}^e(n + 1) = 0,$$

$$\phi_i^e(n + 1) = \phi_{i+1}^e(n + 1) + \phi_i^*(n), \qquad i = k, k - 1, \ldots, 1;$$

E   Evaluating $f_{n+1}^p = f(x_{n+1}, p_{n+1})$;

C   Correcting $y_{n+1} = p_{n+1} + h_{n+1}g_{k+1,1}(f_{n+1}^p - \phi_1^e(n + 1))$;

E   Evaluating $f_{n+1} = f(x_{n+1}, y_{n+1})$,

$$\phi_{k+1}(n + 1) = f_{n+1} - \phi_1^e(n + 1).$$

$$\phi_i(n + 1) = \phi_i^e(n + 1) + \phi_{k+1}(n + 1), \qquad i = k, k - 1, \ldots, 1.$$

Notice that advancing the differences is very cheap though there may be a lot of them. Much of the overhead results from the computation of the $g_{i,1}$.

We stated at the beginning of this chapter that our code would take advantage of steps taken with a constant size. Let us go into this matter now and examine the programming of the computation.

Let the variable n_s be the number of successive steps taken with constant step size h, including the current one. When $\psi_i(n + 1)$ is computed it is written over the stored value of $\psi_i(n)$. This is economical of storage but, more importantly, it saves a lot of work. From the basic definitions

$$\psi_i(n + 1) = \psi_i(n) = ih \qquad \text{for } i = 1, \ldots, n_s - 1,$$

so to compute the $\psi_i(n + 1)$ we need only start with the value

$$\psi_{n_s}(n + 1) = n_s h$$

and then obtain the remaining elements by the relation

$$\psi_i(n + 1) = \psi_{i-1}(n) + h \qquad i = n_s + 1, \ldots, k.$$

In the same way we find that

$$\alpha_i(n + 1) = \alpha_i(n)$$
$$\beta_i(n + 1) = \beta_i(n), \qquad i = 1, \ldots, n_s - 1$$

and that these arrays can be advanced from n to $n + 1$ by computing only

$$\alpha_{n_s}(n + 1) = \frac{1}{n_s},$$

$$\alpha_i(n + 1) = \frac{h}{\psi_i(n + 1)} \qquad i = n_s + 1, \ldots, k,$$

$$\beta_{n_s}(n + 1) = 1,$$

$$\beta_i(n + 1) = \beta_{i-1}(n + 1) \frac{\psi_{i-1}(n + 1)}{\psi_{i-1}(n)} \qquad i = n_s + 1, \ldots, k.$$

Obviously the closer we are to working with a constant step size, the less work need be done in computing these quantities.

More significant savings are realized in the generation of the $g_{i,1}$ when the step size is constant. To explain clearly how this is done we shall describe some notation which, although it is standard, may still be unfamiliar to the reader. Let the vector (one dimensional array) $\mathbf{V}$ have elements $\mathbf{V}(1), \mathbf{V}(2), \ldots$. The expression $\mathbf{V}(2)$ refers both to the storage location in the computer memory for this element of the array, and to the number stored there. If we write $\mathbf{V}(2) = 3$, we mean that the number 3 is stored in the memory location allocated to the second element of the array $\mathbf{V}$. If we write $\mathbf{V}(1) + \mathbf{V}(2)$, we mean to add the numbers stored in these two locations. The replacement operator "$\leftarrow$" is used in expressions such as $\mathbf{V}(2) \leftarrow 3$, and $\mathbf{V}(2) \leftarrow \mathbf{V}(1) + \mathbf{V}(2)$, which mean to replace the contents of the location $\mathbf{V}(2)$ (overwrite it) with the number 3, and, respectively, with the sum of the contents of $\mathbf{V}(1)$ and $\mathbf{V}(2)$.

In the array

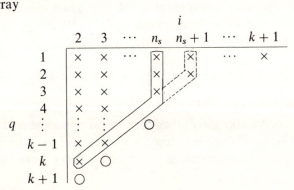

we generate the entries using the equations of (10):

$$g_{2,q} = \frac{1}{q(q+1)},$$

and

$$g_{i,q} = g_{i-1,q} - \alpha_{i-1}(n+1)g_{i-1,q+1}.$$

But $\alpha_{i-1}(n+1) = \alpha_{i-1}(n)$ for $i \leq n_s$, so that columns $i = 2$ through $i = n_s$ are unchanged from the previous step; we need only generate the columns $n_s + 1$ through $k + 1$. It is not necessary to store the triangular array since most elements in the table are not used in generating $g_{n_s+1,q}$. Let the vector $\mathbf{V}(q)$ hold the elements encircled by the solid line in the schematic, namely,

$$\mathbf{V}(k) = g_{2,k}$$
$$\mathbf{V}(k-1) = g_{3,k-1}$$
$$\vdots$$
$$\mathbf{V}(k+2-n_s) = g_{n_s,k+2-n_s}$$
$$\mathbf{V}(k+1-n_s) = g_{n_s,k+1-n_s}$$
$$\vdots$$
$$\mathbf{V}(2) = g_{n_s,2}$$
$$\mathbf{V}(1) = g_{n_s,1}.$$

We update $\mathbf{V}(q)$ for this step by forming and overwriting

$$g_{n_s+1,q} = \mathbf{V}(q) \leftarrow \mathbf{V}(q) - \alpha_{n_s}(n+1)\mathbf{V}(q+1) \qquad q = 1, 2, \ldots, k+1-n_s,$$

that is, $\mathbf{V}$ is extended to hold column $n_s + 1$. These new elements, encircled by the dashed line, are copied into a working vector $\mathbf{W}$, and $g_{n_s+1,1}$ is stored. The remaining coefficients $g_{i,1}$ are generated in $\mathbf{W}$ by

$$g_{i,q} = \mathbf{W}(q) \leftarrow \mathbf{W}(q) - \alpha_{i-1}(n+1)\mathbf{W}(q+1) \qquad q = 1, 2, \ldots, k+2-i$$

and storing elsewhere $g_{i,1} = \mathbf{W}(1)$ for each $i = n_s + 2, \ldots, k+1$.

Changes in order take place; they correspond to adding or dropping an interpolation point. If there is a step size change and an order change, no special account of the order change is required; a step change forces all the coefficients to be completely recomputed. If the step is not changed and the order is reduced, the only effect is that terms like $\alpha_k(n+1)$ are simply not used; thus no special action is required. If the step is not changed and the order is raised, the stored values are not affected but additional terms must be calculated. For the α_i, β_i, and ψ_i this is automatically done when the code compares n_s to the order. However, another diagonal must be added to the $g_{i,q}$ table. For $k = k_{new} = k_{old} + 1$, one can add the known value $g_{2,k}$ as $V(k)$ and then update that part of $V(q)$ holding the diagonal by

$$V(k-j) \leftarrow V(k-j) - \alpha_{j+1}(n+1)V(k-j+1) \qquad j = 1, 2, \ldots, n_s - 2.$$

This is represented by circles in the table. After updating the diagonal the remainder of the computation is just as before.

In this way the integration coefficients are efficiently generated with modest storage requirements. By proceeding in this manner the fact that $n_s - 1$ previous steps of the same size have been taken means that $g_{1,1}, g_{2,1}, \ldots, g_{n_s-1,1}$ do not have to be computed. As noted earlier the computation of $g_{1,1}, \ldots, g_{k+1,1}$ is a large part of the overhead, so this represents a very considerable reduction in cost. Another related saving occurs in the prediction. Because $\phi_i^*(n)$ is written over $\phi_i(n)$ and $\phi_i^*(n) = \beta_i(n+1)\phi_i(n) = \phi_i(n)$ for $i = 1, \ldots, n_s$, we need only compute $\phi_i^*(n)$ for $i = n_s + 1, \ldots, k$. This reduces the overhead substantially when a large number of equations is being integrated since a set of differences is formed for each unknown.

We turn now to the task of interpolating the solution. The code chooses the step size to be as large as possible while still meeting a given error tolerance, so it rarely hits an output point using this step. To force the code to use smaller steps would be inefficient, hence we ordinarily make the code integrate beyond the given point and then interpolate to get the required value. We showed in Chapter 4 that interpolating in this way gives approximations as accurate as those found at mesh points. Sometimes it is not permissible to go beyond the output point, e.g., when some function becomes undefined there. In such a situation the step size should be limited so as to land at the output point.

Let x_{out} be an output point such that

$$x_n \leq x_{out} \leq x_{n+1}$$

and let $P_{k+1,n+1}(x)$ satisfy

$$P_{k+1,n+1}(x_{n+2-j}) = f_{n+2-j} \qquad j = 1, 2, \ldots, k+1.$$

If we define

$$h_I = x_{\text{out}} - x_{n+1},$$

$$s = \frac{(x - x_{n+1})}{h_I}$$

and use the other definitions of (4), then a typical term of $P_{k+1,n+1}(x)$,

$$(x - x_{n+1})(x - x_n)\cdots(x - x_{n-i+3})f[x_{n+1}, x_n, \ldots, x_{n-i+2}],$$

may be written

$$\left(\frac{sh_I}{\psi_1(n+1)}\right) \cdot \left(\frac{sh_I + \psi_1(n+1)}{\psi_2(n+1)}\right) \cdots \left(\frac{sh_I + \psi_{i-2}(n+1)}{\psi_{i-1}(n+1)}\right)\phi_i(n+1).$$

This is already different from the form used in the prediction because the use of the $\psi_j(n+1)$ in the numerators instead of the $\psi_j(n+2)$ means that we have $\phi_i(n+1)$ instead of $\phi_i^*(n+1)$. As we did with the prediction, we condense the notation by introducing $c_{i,n+1}^I(s)$,

$$c_{i,n+1}^I(s) = \begin{cases} 1 & i = 1, \\[2ex] \dfrac{sh_I}{\psi_1(n+1)} & i = 2, \\[2ex] \left(\dfrac{sh_I}{\psi_1(n+1)}\right) \cdot \left(\dfrac{sh_I + \psi_1(n+1)}{\psi_2(n+1)}\right) \cdots \left(\dfrac{sh_I + \psi_{i-2}(n+1)}{\psi_{i-1}(n+1)}\right) & i \geq 3. \end{cases}$$

It follows immediately that the solution value at x_{out} is

$$y_{\text{out}} = y_{n+1} + h_I \sum_{i=1}^{k+1} \phi_i(n+1) \int_0^1 c_{i,n+1}^I(s)\, ds.$$

The scheme for evaluating these integrals is derived in a manner similar to that used to derive the predictor but the scheme itself is rather different. Mathematically, the other scheme is equivalent except when $x_{\text{out}} = x_n$, for then $\psi_2(n+2) = 0$ and $\alpha_2(n+2)$ is not defined. For a value of x_{out} very near x_n, numerical difficulties (such as divide checks and overflows) can be expected with the predictor scheme. The new scheme avoids such troubles— but at the expense of computing all the coefficients each time we interpolate.

Introducing

$$\Gamma_i(s) = \begin{cases} \dfrac{sh_I}{\psi_1(n+1)} & i = 1, \\[3mm] \dfrac{sh_I + \psi_{i-1}(n+1)}{\psi_i(n+1)} & i \geq 2, \end{cases}$$

we may write for $i > 1$

$$c^I_{i,n+1}(s) = c^I_{i-1,n+1}(s) \cdot \Gamma_{i-1}(s).$$

Repeated integration by parts shows

$$c^{I(-q)}_{i,n+1}(s) = \Gamma_{i-1}(s) c^{I(-q)}_{i-1,n+1}(s) - \frac{q \cdot h_I}{\psi_{i-1}(n+1)} c^{I(-q-1)}_{i-1,n+1}(s).$$

In terms of the scaled quantities

$$g^I_{i,q} = (q-1)! \, c^{I(-q)}_{i,n+1}(1)$$

we have the recursion

$$g^I_{i,q} = \begin{cases} \dfrac{1}{q} & i = 1, \\[3mm] \Gamma_{i-1}(1) \cdot g^I_{i-1,q} - \eta_{i-1} g^I_{i-1,q+1} & i \geq 2 \end{cases}$$

where

$$\eta_i = \frac{h_I}{\psi_i(n+1)}.$$

The interpolated solution value is

$$y_{\text{out}} = y_{n+1} + h_I \sum_{i=1}^{k+1} g^I_{i,1} \phi_i(n+1).$$

Although this requires that the coefficients $g^I_{i,1}$ be computed for each output point, no extra function evaluations are required and the step size selection is not impaired by output points.

Some applications also require the derivative of the approximate solution at x_{out}. This is easy to obtain since we are working with $P_{k+1,n+1}(x)$ and $y'_{\text{out}} = P_{k+1,n+1}'(x_{\text{out}})$. The derivation of an algorithm is left as an exercise.

EXERCISE 6

Show that

$$y'_{\text{out}} = \sum_{i=1}^{k+1} \rho_i \, \phi_i(n+1),$$

where the $\rho_i = c^t_{i,n+1}(1)$ can be computed by

$$\rho_1 = 1,$$

$$\rho_i = \rho_{i-1} \cdot \Gamma_{i-1}(1) \qquad i = 2,\ldots, k+1.$$

6

ERROR ESTIMATION
AND CONTROL

In this chapter and the next we discuss the estimation of the errors that are being made during an integration and the adjustment of step size and order to control these errors while still solving the problem efficiently. There is a great deal of art in doing this since the theory available is fragmentary or somewhat unrealistic. This chapter is concerned with theoretical aspects of error estimation and control; the next combines these results with heuristic arguments and experience to develop the actual algorithms.

The convergence results derived in Chapter 4 permit nearly any kind of variation of both the step size and order, but since no assumptions are made about an effective choice of step size and order, they may not be relevant to actual computation. The results obtained assuming a fixed order are often realistic because in many cases it is true that the order k is constant for a large number of steps. When the order is varied, the best conclusion that can be drawn, from the viewpoint of Chapter 4, is that convergence occurs at the rate of the lowest order formula. But because the codes start with $k = 1$, this viewpoint gives a completely false impression. A realistic convergence result can be inferred from Theorem 4 of Chapter 4 if we assume that the step size and order are so chosen at each step that the local truncation error δ_n satisfies $|\delta_n| \leq h_{n+1}\varepsilon$ for some user-supplied tolerance ε. The theorem then states that the global error is bounded by

$$|y(x_n) - y_n| \le \frac{\varepsilon}{L(\alpha^* + HLC\alpha)} \exp DL(\alpha^* + HLC\alpha).$$

According to this bound the accuracy is not affected by changes of order. Variation of the order serves two purposes. One is to start the code and the other is to achieve the required accuracy at each step as efficiently as possible.

Control of the local truncation error shows that the conceptual difficulty of varying the order in a limit analysis can be surmounted. However, we shall control a closely related quantity which is connected to the numerical values in a more fundamental way. And so we shall begin by considering carefully just what the numerical methods attempt to do at each step. We have emphasized the analogy between the Taylor series and Adams methods and we shall rely upon it again for guidance in the matters of error estimation and control.

Let us define $u_n(x)$ as the solution of

$$u'_n(x) = f(x, u_n(x)),$$

$$u_n(x_n) = y_n.$$

The Taylor series method of order k takes

$$y_{n+1} = y_n + hu'_n(x_n) + \frac{h^2}{2!} u''_n(x_n) + \cdots + \frac{h^k}{k!} u_n^{(k)}(x_n).$$

More generally it approximates $u_n(x)$ for all x in the interval $[x_n, x_n + h]$ by

$$y_I(x) = y_n + (x - x_n)u'_n(x_n) + \cdots + \frac{(x - x_n)^k}{k!} u_n^{(k)}(x_n).$$

Because the basic aim of the method is to approximate the local solution $u_n(x)$, rather than the global solution $y(x)$, local truncation error is not a natural quantity to control. A more natural quantity to be controlled is what we call the local error,

$$LE_{n+1} = u_n(x_{n+1}) - y_{n+1}.$$

This concept of local error is useful for any method which proceeds from an approximate solution value y_n at x_n to the value y_{n+1} at x_{n+1}. In Chapter 1 we discussed a very natural estimate of the local error for the

Taylor series method:

$$LE_{n+1} = \frac{h^{k+1}u_n^{(k+1)}(\xi)}{(k+1)!} \doteq \frac{h^{k+1}u_n^{(k+1)}(x_n)}{(k+1)!}.$$

In the same way we find that the error at any point in the interval is

$$u_n(x) - y_I(x) = \frac{(x-x_n)^{k+1}u_n^{(k+1)}(\eta)}{(k+1)!} \doteq \frac{(x-x_n)^{k+1}u_n^{(k+1)}(x_n)}{(k+1)!}.$$

Since

$$\left| \frac{(x-x_n)^{k+1}u_n^{(k+1)}(x_n)}{(k+1)!} \right| \leq \left| \frac{h^{k+1}u_n^{(k+1)}(x_n)}{(k+1)!} \right|,$$

we *estimate* that the largest error occurs at the end of the interval. Control of the local error means that we try to approximate the local solution uniformly well over the entire interval $[x_n, x_{n+1}]$.

The intent of the Adams methods, as expressed in their derivation in Chapter 3, is also to approximate the local solution throughout $[x_n, x_{n+1}]$. It is illuminating to look at the matter from a slightly different viewpoint. In Chapter 4 we defined the function $y_I(x)$ as the approximation to $y(x)$ on the entire interval $[a, b]$. The function $f(x, y_I(x))$ has the values $f(x_i, y_I(x_i)) = f(x_i, y_i)$ at the mesh points. On the interval $[x_n, x_{n+1}]$ the function $y_I'(x)$ is a polynomial which, if the step were taken at order k, would interpolate to the values f_{n+1-j} for $j = 0, 1, \ldots, k$. Thus

$$y_I'(x) = f(x, y_I(x)) + r(x) \qquad x_n < x \leq x_{n+1},$$

where the residual, $r(x)$, is the error of interpolation. In this way we see that the Adams methods lead to an approximate solution $y_I(x)$ which is the exact solution of an equation "close" to the one given. If the residual is small and the problem given is well posed, Theorem 2 of Chapter 1 says $y_I(x)$ is close to $y(x)$. Just how close it is depends on how stable the problem is.

Now that the local error is recognized as a natural quantity for us to attempt to estimate and control, we must ask if its control leads to a conclusion like that drawn for the control of local truncation error. We say that the local error is controlled per unit step, that is, controlled relative to the length of the step, if at each step

$$|LE_{n+1}| \leq h_{n+1}\varepsilon.$$

We shall also say it is controlled by the criterion of error per unit step in a generalized sense if, for some (unknown) constant σ,

$$|\text{LE}_{n+1}| \le \sigma h_{n+1}\varepsilon \tag{1}$$

at each step.

The global error e_{n+1} can be split into

$$e_{n+1} = y(x_{n+1}) - y_{n+1} = y(x_{n+1}) - u_n(x_{n+1}) + \text{LE}_{n+1}.$$

The functions $y(x)$ and $u_n(x)$ are solutions of the same differential equation with different initial values at x_n. The equation is Lipschitzian, so Theorem 2 of Chapter 1 implies that

$$|y(x) - u_n(x)| \le |y(x_n) - u_n(x_n)| \exp L|x - x_n|.$$

Thus, we can see that

$$|e_{n+1}| \le |e_n| \exp Lh_{n+1} + \sigma h_{n+1}\varepsilon.$$

Since the initial error is zero, repetition of this inequality for $n = 0, 1, \ldots,$ $m - 1$ leads to

$$|e_m| \le \sum_{n=0}^{m-1} \sigma h_{n+1}\varepsilon \exp Lh_{n+1}$$

$$\le \sigma\varepsilon \exp L(x_m - a) \sum_{n=0}^{m-1} h_{n+1}.$$

Consequently

$$|y(x_m) - y_m| \le \varepsilon\sigma(x_m - a)\exp L(x_m - a),$$

which is the same kind of result we found for control of the local truncation error.

A general procedure for estimating local errors is to compare the results of formulas of different orders. Let $y_{n+1}(k)$ be the result of stepping to x_{n+1} with a method of order k. To be specific, suppose we use the Taylor series method. If we assume on heuristic grounds that $y_{n+1}(k + 1)$ is more accurate than $y_{n+1}(k)$, which is to say that it is closer to $u_n(x_{n+1})$, then

$$y_{n+1}(k + 1) - y_{n+1}(k) = [u_n(x_{n+1}) - y_{n+1}(k)] - [u_n(x_{n+1}) - y_{n+1}(k + 1)]$$

$$\doteq u_n(x_{n+1}) - y_{n+1}(k).$$

In this way we can estimate the local error associated with the less accurate result. For the Taylor series method we have

$$y_{n+1}(k+1) - y_{n+1}(k) = \frac{h^{k+1}}{(k+1)!} u_n^{(k+1)}(x_n)$$

which is the same estimate arrived at earlier in another way.

When at order k, we compute $y_{n+1}(k)$, then estimate the local error LE_{n+1}, and finally make the step and order decisions using this estimate. What we really hope to compute in the step is $u_n(x_{n+1}) = y_{n+1}(k) + \text{LE}_{n+1}$. If we believe the local error estimate to be reasonably accurate, it would seem desirable to use it to improve the accuracy of $y_{n+1}(k)$. So, we add in the estimated error to form

$$y_{n+1}(k) + (y_{n+1}(k+1) - y_{n+1}(k)) = y_{n+1}(k+1);$$

this is called "local extrapolation." Formally the result is obtained by using a higher order formula, so for "small" step sizes extrapolation increases the accuracy. An important question when we apply local extrapolation to the Adams methods is, "What happens with the large step sizes?" We shall respond to this later. We shall always do local extrapolation. Since the step size and order decisions are made with respect to a formula of order k, we refer to the method as being of order k even though the numerical values that are accepted come from a formula of order $k + 1$.

When we do local extrapolation, there are two approximate solutions hence two local errors to be considered. Suppose we are using the Taylor series method of order k. Then the local error, le, used in selecting the step size is

$$\text{le}_{n+1} = u_n(x_{n+1}) - y_{n+1}(k)$$

$$\doteq y_{n+1}(k+1) - y_{n+1}(k).$$

The final value accepted is

$$y_{n+1}(k+1) = y_{n+1}(k) + (y_{n+1}(k+1) - y_{n+1}(k))$$

and thus the error actually made in the step is

$$\text{LE}_{n+1} = u_n(x_{n+1}) - y_{n+1}(k+1).$$

We shall reserve the capital letters to indicate errors actually made, in agreement with our earlier analysis, and lower case letters to indicate other errors being estimated. It seems silly to discard $y_{n+1}(k+1)$ if we must compute it anyway, so local extrapolation is appealing; but it leads to a complication. How do we choose the step size so as to control the local error actually made by a generalized criterion of error per unit step? It turns out that we should require

$$|le_{n+1}| \leq \varepsilon.$$

Without considering all the details in this chapter the essence, for the Taylor series method, is that for the sake of efficiency we choose the largest step size which appears to satisfy the criterion. Thus

$$|le_{n+1}| \doteq \left| \frac{h_{n+1}^{k+1} u_n^{(k+1)}(x_n)}{(k+1)!} \right| \doteq \varepsilon.$$

But then

$$|LE_{n+1}| = \left| \frac{h_{n+1}^{k+2} u_n^{(k+2)}(\zeta)}{(k+2)!} \right| \doteq \varepsilon h_{n+1} \left| \frac{u_n^{(k+2)}(\zeta)}{(k+2) u_n^{(k+1)}(x_n)} \right| \leq \sigma \varepsilon h_{n+1}$$

for a suitable constant σ. And so the resulting local error that is actually made is controlled as in (1).

For our further discussion of the Adams methods we need to introduce some additional notation. It will be necessary to estimate the errors that would have been made had a step been carried out using any one of several orders, so that an efficient order can be chosen. In addition there are various predicted and corrected values which must be kept straight. The result of predicting the solution at x_{n+1} using the formula of order $k-1$ is denoted by $p_{n+1}(k-1)$ and similarly for other orders. Because the analysis and computation are in terms of $p_{n+1}(k)$ we shall distinguish this value by writing simply p_{n+1} and, correspondingly, the notation f_{n+1}^p will be used for $f(x_{n+1}, p_{n+1})$. The result of correcting a predicted value of, say, order $k-1$ with a corrector of the same order will be denoted by $y_{n+1}(k-1)$. Local errors like $u_n(x_{n+1}) - y_{n+1}(k-1)$ will be written $le_{n+1}(k-1)$. The primary object of the analysis is $y_{n+1}(k)$ but the computations are based on the result of using p_{n+1} and the corrector of order $k+1$, which we denote by y_{n+1}.

So far in this text we have been mainly concerned with the procedure using a predictor of order k and a corrector of order $k+1$ to generate y_{n+1}. In parallel we have developed the procedure using both predictor and corrector of order k to generate $y_{n+1}(k)$. We saw in Chapter 4 that the local truncation errors of the two procedures are of orders $k + 2$ and $k + 1$, respectively. This is in agreement with our expectation on heuristic grounds, discussed in Chapters 2 and 3, that y_{n+1} is more accurate than $y_{n+1}(k)$. The limit analysis of Chapter 4 says that for "small" step sizes y_{n+1} is indeed more accurate. For these reasons we anticipate that the general procedure for constructing error estimators will be successful. The estimator is

$$\mathrm{le}_{n+1}(k) = u_n(x_{n+1}) - y_{n+1}(k) \doteq y_{n+1} - y_{n+1}(k).$$

This estimator is very practical since the formulas of Chapter 5,

$$y_{n+1} = p_{n+1} + h_{n+1} g_{k+1,1} \, \phi_{k+1}^p(n+1),$$

$$y_{n+1}(k) = p_{n+1} + h_{n+1} g_{k,1} \, \phi_{k+1}^p(n+1),$$

lead to

$$\mathrm{le}_{n+1}(k) \doteq h_{n+1}(g_{k+1,1} - g_{k,1})\phi_{k+1}^p(n+1).$$

In Chapter 8 we consider the effect of extrapolation on an important stability property of the computational procedure which is relevant to "large" step sizes. There we find that local extrapolation is considerably better than not extrapolating. Because of this and because it is more accurate for "small" step sizes we use y_{n+1}, which represents the result of $p_{n+1}(k)$ and the corrector of order $k + 1$. We prefer to regard y_{n+1} as arising from $y_{n+1}(k)$ by local extrapolation, whence the term "PECE procedure of order k."

The model situation for studying error estimation is the same as that for the convergence results of Chapter 4. A sequence of integrations is considered with the maximum step size H tending to zero and the step sizes satisfying the constraints of Theorem 3 of Chapter 4. Starting values and a choice of orders are made so that the global errors are uniformly $O(H^{k+1})$, e.g. the PECE procedure of order k is always used. The basic questions are: What is the local error in stepping to x_{n+1} if a predictor and corrector of order k are used? What happens if orders $k - 2, k - 1, k + 1$ are used? Can these errors be estimated using only information available from actually taking the step with the predictor of order k and corrector of order $k + 1$?

As in most of our arguments about convergence we need to consider first the error of the predicted value. From their definitions we find

$$u_n(x_{n+1}) - p_{n+1} = \int_{x_n}^{x_{n+1}} u_n'(t) - P_{k,n}(t)\, dt.$$

Let $P_{k,n}(x)$ be the polynomial of degree $k-1$ satisfying

$$P_{k,n}(x_{n+1-j}) = u'_n(x_{n+1-j}) \qquad j = 1, 2, \ldots, k.$$

Using this polynomial we can split the local error of the predictor into two parts:

$$u_n(x_{n+1}) - p_{n+1} = \int_{x_n}^{x_{n+1}} u'_n(t) - P_{k,n}(t) \, dt + \int_{x_n}^{x_{n+1}} P_{k,n}(t) - P_{k,n}(t) \, dt. \quad (2)$$

The first integral is the error due to approximating $u'_n(t)$ by an interpolating polynomial. It will be called "the local discretization error." We discussed this error in Chapter 3 where we were approximating $y'(t)$ rather than a local solution $u'_n(t)$. It was then called "local truncation error." The assessment of this term is of course the same as that of Chapter 3 and leads to the conclusion that the local discretization error of the Adams-Bashforth formula is $O(H^{k+1})$.

The second integral in (2) is the error due to inaccurate memorized values y_{n+1-j}. The error induced by inaccurate memorized values, as represented by this integral, is best examined in the Lagrangian form

$$\int_{x_n}^{x_{n+1}} P_{k,n}(t) - P_{k,n}(t) \, dt = h_{n+1} \sum_{j=1}^{k} \alpha_{k,j} [f(x_{n+1-j}, u_n(x_{n+1-j}))$$

$$- f(x_{n+1-j}, y_{n+1-j})].$$

The integral can be bounded using the Lipschitz condition,

$$\left| \int_{x_n}^{x_{n+1}} P_{k,n}(t) - P_{k,n}(t) \, dt \right| \le HL \sum_{j=1}^{k} |\alpha_{k,j}| \, |u_n(x_{n+1-j}) - y_{n+1-j}|. \quad (3)$$

In Chapter 4 we argued that because of the bounds on the ratio of successive step sizes the sum $\sum |\alpha_{k,j}|$ is uniformly bounded. The discrepancy between the local solution and the computed solution can be split and bounded by

$$|u_n(x_{n+1-j}) - y_{n+1-j}| \le |u_n(x_{n+1-j}) - y(x_{n+1-j})| + |y(x_{n+1-j}) - y_{n+1-j}|. \quad (4)$$

The second term in (4) is $O(H^{k+1})$ by the assumption about convergence. Now $u_n(x)$ and $y(x)$ are solutions of the same differential equation with different initial values at x_n. The equation is well posed and by Theorem 2 of Chapter 1 the solutions satisfy

$$|u_n(x) - y(x)| \le |u_n(x_n) - y(x_n)| \exp L|x - x_n|$$

hence

$$|u_n(x_{n+1-j}) - y(x_{n+1-j})| \le |y_n - y(x_n)| \exp LkH \qquad j = 1, \ldots, k.$$

This bound for the first term in (4) and convergence complete the proof that locally $(x_{n+1-k} \le x \le x_{n+1}$, for instance) the solution $y(x)$ and the local solution $u_n(x)$ agree to terms of order H^{k+1}. These results, along with the bound (3), show that the effect of inaccurate memorized values is $O(H^{k+2})$; and finally

$$u_n(x_{n+1}) - p_{n+1} = O(H^{k+1}).$$

It is now easy to find the order of the local error of the predictor-corrector process. We have

$$le_{n+1}(k) = u_n(x_{n+1}) - y_{n+1}(k) = \int_{x_n}^{x_{n+1}} u_n'(t) - P_{k,n}^*(t) \, dt.$$

Using the polynomial $P_{k,n+1}(x)$ of degree $k-1$ satisfying

$$P_{k,n+1}(x_{n+1-j}) = u_n'(x_{n+1-j}) \qquad j = 0, 1, \ldots, k-1,$$

we split the local error into the discretization error of the corrector formula and the error due to the memorized and predicted values:

$$le_{n+1}(k) = \int_{x_n}^{x_{n+1}} u_n'(t) - P_{k,n+1}(t) \, dt + \int_{x_n}^{x_{n+1}} P_{k,n+1}(t) - P_{k,n}^*(t) \, dt.$$

Just as before, the local discretization error of the corrector formula is $O(H^{k+1})$. The second integral is

$$\int_{x_n}^{x_{n+1}} P_{k,n+1}(t) - P_{k,n}^*(t) \, dt = h_{n+1} \sum_{j=1}^{k-1} \alpha_{k,j}^*[f(x_{n+1-j}, u_n(x_{n+1-j}))$$

$$- f(x_{n+1-j}, y_{n+1-j})]$$

$$+ h_{n+1}\alpha_{k,0}^*[f(x_{n+1}, u_n(x_{n+1}))$$

$$- f(x_{n+1}, p_{n+1})].$$

Using the result about the local error of the predicted value we argue as before that this integral is $O(H^{k+2})$. Thus we find $le_{n+1}(k)$ is $O(H^{k+1})$. An important point to be noted is that with the assumptions made, the local discretization error dominates the error due to the memorized and predicted values. An entirely similar argument for the corrector of order $k+1$ has the discretization error and the error due to inaccurate approximate solutions both $O(H^{k+2})$. But then

$$le_{n+1}(k) = u_n(x_{n+1}) - y_{n+1}(k)$$

$$= y_{n+1} - y_{n+1}(k) + u_n(x_{n+1}) - y_{n+1}$$

$$= y_{n+1} - y_{n+1}(k) + O(H^{k+2}).$$

This result justifies the estimate we suggested earlier on heuristic grounds.

The local error that would have been made had the step been taken with a predictor and corrector of order $k - 1$ can be analyzed just as was done for order k. It follows that $\text{le}_{n+1}(k - 1)$ is $O(H^k)$ and that if we had also computed $y_{n+1}(k)$, we could estimate $\text{le}_{n+1}(k - 1)$ by

$$\text{le}_{n+1}(k - 1) = u_n(x_{n+1}) - y_{n+1}(k - 1)$$

$$= y_{n+1}(k) - y_{n+1}(k - 1) + u_n(x_{n+1}) - y_{n+1}(k)$$

$$= y_{n+1}(k) - y_{n+1}(k - 1) + O(H^{k+1}).$$

As in the estimate of $\text{le}_{n+1}(k)$, the important thing is to have a more accurate approximate solution available. As far as computation is concerned, the crucial question is whether or not we can estimate $\text{le}_{n+1}(k - 1)$ without actually computing $y_{n+1}(k - 1)$. This quantity is

$$y_{n+1}(k - 1) = y_n + h_{n+1} \sum_{j=1}^{k-2} \alpha^*_{k-1,j} f_{n+1-j} + h_{n+1} \alpha^*_{k-1,0} f(x_{n+1}, p_{n+1}(k - 1)).$$

The thing we wish to avoid is the evaluation of $f(x_{n+1}, p_{n+1}(k - 1))$. Can we substitute the more accurate value f^p_{n+1} that is available from the computation of $y_{n+1}(k)$? To see that this is possible let us first define

$$\bar{y}_{n+1}(k - 1) = y_n + h_{n+1} \sum_{j=1}^{k-2} \alpha^*_{k-1,j} f_{n+1-j} + h_{n+1} \alpha^*_{k-1,0} f^p_{n+1}$$

$$= y_n + h_{n+1} \sum_{i=1}^{k-1} (g_{i,1} - g_{i-1,1}) \phi^p_i(n + 1).$$

(The equivalence of these forms is shown in Exercise 5 of Chapter 5.) Then

$$|\bar{y}_{n+1}(k - 1) - y_{n+1}(k - 1)| = |h_{n+1} \alpha^*_{k-1,0}[f(x_{n+1}, p_{n+1})$$

$$- f(x_{n+1}, p_{n+1}(k - 1))]|$$

$$\leq h_{n+1} |\alpha^*_{k-1,0}| L |p_{n+1} - p_{n+1}(k - 1)|$$

$$\leq h_{n+1} |\alpha^*_{k-1,0}| L\{|p_{n+1} - u_n(x_{n+1})|$$

$$+ |u_n(x_{n+1}) - p_{n+1}(k - 1)|\}$$

$$= O(H^{k+1}).$$

Thus

$$\text{le}_{n+1}(k - 1) = y_{n+1}(k) - y_{n+1}(k - 1) + O(H^{k+1})$$

$$= y_{n+1}(k) - \bar{y}_{n+1}(k - 1) + O(H^{k+1})$$

$$= h_{n+1}(g_{k,1} - g_{k-1,1}) \phi^p_k(n + 1) + O(H^{k+1}).$$

In exactly the same way we can obtain computable estimates of the local error had the step been taken with any order lower than k. We shall also use

$$\text{le}_{n+1}(k-2) = h_{n+1}(g_{k-1,1} - g_{k-2,1})\phi^p_{k-1}(n+1) + O(H^k)$$

in the algorithm for choosing the order. Not only can these local error estimates be obtained without taking the step at these orders but they are also available before the final evaluation for the PECE formulas of order k. Thus we compute through PEC and see if the test $|\text{le}_{n+1}(k)| \leq \varepsilon$ for a successful step is passed. If the step is unsuccessful, we do not make the final evaluation. Even so, we have available estimates of $\text{le}_{n+1}(k)$, $\text{le}_{n+1}(k-1)$, and $\text{le}_{n+1}(k-2)$ so that lowering the order can be considered when the step is repeated.

In all these cases the inaccurate memorized values lead to errors of a higher order than the local discretization error. As a consequence they can be neglected. When this is done, there is no essential difference between the Adams methods with memory and the Taylor series methods without memory. Estimating $\text{le}_{n+1}(k+1)$ is more delicate because the memorized values appear to lead to errors of the same order as the local discretization error. An additional hypothesis and a more careful examination of the contribution due to the memorized values again show the local discretization error dominates, provided that local extrapolation is done.

So far we have not had to make any assumption about the relationship between the step sizes in the successive integrations which are part of the model for studying convergence. Now we shall assume that a constant step size and a fixed order k are used. Just how realistic this is and the implications thereof will be discussed later. At the end of Chapter 4 we proved with these same assumptions that

$$y(x_{n+1-j}) = y_{n+1-j} + h^{k+1}\phi(x_{n+1-j}) + O(h^{k+2}),$$

for a smooth function $\phi(x)$. Up to a point the analysis for $\text{le}_{n+1}(k+1)$ is like that for $\text{le}_{n+1}(k)$. We find that

$$u_n(x_{n+1}) - p_{n+1}(k+1) = O(h^{k+2})$$

noting, however, that in this case the local discretization error is of the same order as the error due to inaccurate memorized values. The error due to the memorized and predicted values in the corrector formula is bounded by

$$\left| \int_{x_n}^{x_{n+1}} P_{k+1,n+1}(t) - P^*_{k+1,n}(t)\, dt \right| \leq hL \sum_{j=1}^{k} |\alpha^*_{k+1,j}| \, |u_n(x_{n+1-j}) - y_{n+1-j}|$$

$$+ hL |\alpha^*_{k+1,0}| \, |u_n(x_{n+1}) - p_{n+1}(k+1)|.$$

We have already used the fact that asymptotically there is little difference between the local solution $u_n(x)$ and the global solution $y(x)$. In Chapter 1 we referred to the equation of first variation, which provides more quantitative information about the relation between these two solutions. Using it we find that

$$u_n(x) = y(x) + \delta v(x) + O(\delta^2)$$

where

$$\delta = u_n(x_n) - y(x_n) = y_n - y(x_n) = O(h^{k+1})$$

and

$$v' = f_y(x, y(x))v, \qquad v(x_n) = 1.$$

From this we can see that, for $j = 0, 1, \ldots, k$,

$$u_n(x_{n+1-j}) - y_{n+1-j} = (y(x_{n+1-j}) - y_{n+1-j})$$
$$- (y(x_n) - y_n)v(x_{n+1-j}) + O(h^{2k+2})$$
$$= h^{k+1}\phi(x_{n+1-j}) - h^{k+1}\phi(x_n)(1 + O(h)) + O(h^{2k+2})$$
$$= -(j-1)h^{k+2}\phi'(x_n) + O(h^{k+2}) = O(h^{k+2}).$$

We already know that

$$u_n(x_{n+1}) - p_{n+1}(k+1) = O(h^{k+2}),$$

so these facts and the bound on the error due to the memorized and predicted values show that their effect is $O(h^{k+3})$.

These assumptions are not as unrealistic as they might appear when taken out of context. The algorithms of Chapter 7 result in codes which have a strong tendency to take steps in groups of constant size and at a fixed order (especially the former). Within such a group this model is reasonably descriptive of the computation. The formal assumptions and the details of the argument obscure what is happening. The ideas are quite plausible: the local solution, $u_n(x)$, amounts to a translation of the global solution, $y(x)$, so as to agree with y_n at x_n. Assuming that the numerical method yields values such that the global error changes smoothly, $u_n(x)$ ought to be closer to $y_{n-1}, y_{n-2}, \ldots, y_{n+1-k}$ than is $y(x)$. Convergence gives the relation

$$u_n(x_{n+1-j}) = y(x_{n+1-j}) + \delta v(x_{n+1-j}) + O(H^{2k+2})$$
$$= y(x_{n+1-j}) + \delta + O(H^{k+2})$$

locally, that is, for $j = 0, 1, \ldots, k$. If we presume that the global error changes smoothly, so that

$$y(x_{n+1-j}) - y_{n+1-j} = y(x_n) - y_n + O(H^{k+2})$$
$$= -\delta + O(H^{k+2}),$$

then it follows that locally

$$u_n(x_{n+1-j}) - y_{n+1-j} = O(H^{k+2}).$$

This is the conclusion we require. Because of this informal, but plausible, argument we can reasonably expect the conclusion to be more realistic than the formal assumptions would suggest.

Attempting to find a practical estimate of $le_{n+1}(k+1)$ exposes another difficulty. If we were to actually take the step with a predictor of order $k+1$ and corrector of order $k+2$ we could make the estimate just as we did for $le_{n+1}(k)$. For constant step size this estimate is

$$le_{n+1}(k+1) = h(\gamma_{k+1} - \gamma_k)\nabla^{k+1}f(x_{n+1}, p_{n+1}(k+1)) + O(h^{k+3}),$$

in an obvious extension of the notation $\nabla^{k+1}f_{n+1}^p$ to remind us which predicted value is being used. It is not possible to substitute f_{n+1}^p for $f(x_{n+1}, p_{n-1}(k+1))$ here since they differ by terms of order $k+1$ which, even when multiplied by the h, are as large as the quantity to be measured. On the other hand, because we extrapolate, f_{n+1} *is* accurate enough. Now from Chapter 2 we know that

$$\nabla^{k+1}f(x_{n+1}, p_{n+1}(k+1)) = f(x_{n+1}, p_{n+1}(k+1))$$
$$+ \sum_{m=1}^{k+1} (-1)^m f_{n+1-m}\binom{k+1}{m}$$
$$= f(x_{n+1}, p_{n+1}(k+1)) - f(x_{n+1}, y_{n+1})$$
$$+ \nabla^{k+1}f_{n+1}$$
$$= \nabla^{k+1}f_{n+1} + O(h^{k+2}).$$

Thus

$$le_{n+1}(k+1) = h(\gamma_{k+1} - \gamma_k)\nabla^{k+1}f_{n+1} + O(h^{k+3})$$
$$= h\gamma_{k+1}^*[\nabla^k f_{n+1} - \nabla^k f_n] + O(h^{k+3}).$$

Even with the additional assumptions we have made, it has proved necessary to extrapolate and to do the final evaluation in order to get a practical estimate of $le_{n+1}(k+1)$.

We have related the local solution $u_n(x)$ to the global solution $y(x)$ by the equation of first variation and the fact that they differ at x_n by the small quantity $\delta = y_n - y(x_n)$. For the smooth functions $f(x, y)$ which concern us, a similar conclusion can be drawn for the derivatives as well [21]:

$$u_n^{(i)}(x_n) = y^{(i)}(x_n) + O(\delta).$$

As a consequence the local discretization error of the corrector is asymptotically equivalent to the local truncation error since we developed (in Chapter 3) the forms

$$\frac{u_n^{(k+1)}(\xi)}{k!} \int_{x_n}^{x_{n+1}} \prod_{j=0}^{k-1} (t - x_{n+1-j}) \, dt$$

$$= \frac{u_n^{(k+1)}(x_n)}{k!} \int_{x_n}^{x_{n+1}} \prod_{j=0}^{k-1} (t - x_{n+1-j}) \, dt + O(H^{k+2}),$$

and

$$\frac{y^{(k+1)}(\eta)}{k!} \int_{x_n}^{x_{n+1}} \prod_{j=0}^{k-1} (t - x_{n+1-j}) \, dt$$

$$= \frac{y^{(k+1)}(x_n)}{k!} \int_{x_n}^{x_{n+1}} \prod_{j=0}^{k-1} (t - x_{n+1-j}) \, dt + O(H^{k+2}),$$

respectively. Thus in the circumstances we have studied, it is true that in the local error the effects of inaccurate memorized and predicted values can be neglected in comparison to the local discretization error. Furthermore, the local discretization error is asymptotically equivalent to the local truncation error. Comparing the results of this chapter to the heuristics of Chapter 3 shows that with a little care in the way the Adams methods are used, the heuristic approach does not lead us astray. In view of the incomplete theory available this is reassuring and augurs well for the successful development of the algorithms in the next chapter.

ORDER AND
STEP SIZE SELECTION

In this chapter we develop algorithms for the selection of step size and order based on the analysis of the preceding chapters. Although the principal aim of the algorithms is to integrate a problem using as few evaluations of the equation as possible, there are other important considerations. Foremost among these are stability, overhead, the validity of the theoretical support, and the detection of problems outside the class for which the codes are intended.

In Chapter 4 we saw that some restrictions on the ratio of successive step sizes are necessary to assure stability in the practical sense discussed there. We saw in Chapter 5 that the overhead is considerably reduced when a constant step size is used. In Chapter 6 we were able to justify an estimate of the local error associated with raising the order only while using a constant step size, though a plausibility argument suggests the estimate of local error is valid in far more general circumstances. We shall see that the selection of an "optimal" step size depends on the step size not varying too much locally. For all these reasons, and others which will appear, we develop algorithms in this chapter which tend to yield steps taken in groups of constant step size, with occasional transitions from one step size to another.

Since the theory of the order and step size algorithms is quite incomplete, there is no best way to proceed. One must consider the fragments of theory available, the remainder of the code, experimentation with variant algorithms, and the experience of other workers in the field. Two codes due to F. T. Krogh [20] and C. W. Gear [3] are largely responsible for the recognition of the power of the variable order, variable step Adams approach to solving differential equations. Experience with these very successful codes and the researches of their authors, e.g. [3, 17, 18], must be considered in developing effective algorithms. Their algorithms and those developed in this chapter differ a great deal in detail, but are broadly the same. The superficial variation is largely due to different emphasis, different ways of achieving the same result, and the fact that the codes are based on different forms of the interpolating polynomial. The main real differences are how changes of step size and how the starting phase of the computation are handled. It is to be hoped that further research will make these algorithms less heuristic, but years of successful computation support an approach along the general lines we now proceed to develop.

ACCEPTING A STEP AND GENERAL CONSIDERATIONS

The first and most important decision is whether or not to accept the result of a step. The selection of a step size and order affects the efficiency of a code but the acceptance criterion determines its reliability. The analysis given in Chapter 6 is quite a reasonable model of real computation and we use the estimate derived there. So we form the quantity

$$\text{ERR} = \left| h_{n+1}(g_{k+1,1} - g_{k,1})\phi^p_{k+1}(n+1) \right| \doteq \left| \text{le}_{n+1}(k) \right|$$

and accept or reject the step depending on whether or not $\text{ERR} \leq \varepsilon$.

The arguments of Chapter 6 pivot on two salient ideas. One is that the discretization error dominates the error due to inaccurate memorized values. The other is a little obscured by the limit arguments being employed. The discretization error arises from the error of interpolation and its estimation is essentially the estimation of the error of an interpolating polynomial. As we saw more clearly in Chapter 2, such an estimate approximates a derivative $f^{(k)}(\xi)/k!$ by a divided difference $f[x_n, x_{n-1}, \ldots, x_{n-k}] = f^{(k)}(\eta)/k!$. How good the estimate is depends principally on how nearly constant the derivative is over the span of the data. Since over this span $f^{(k)}(\zeta) = f^{(k)}(x_n) + O(H)$, the derivative is effectively constant in a limit analysis. In practical computation we, in effect, presume it to be roughly constant. One purpose of the order adjustment algorithm is to help bring this about.

These two ideas and the estimators of the last chapter are the foundation of the algorithms. Let us bring them together to predict the errors that will be seen. On stepping to x_{n+1} we have two possibilities. If the step is unsuccessful, we need to estimate what error would be found if we repeated the step with a smaller step size and possibly a different order. If the step is successful, we need to predict the error that would be made in the next step. In the latter case suppose we were to step to x_{n+2} at order k. The estimate of the local error is

$$\left|\mathrm{le}_{n+2}(k)\right| \doteq \left|h_{n+2}(g_{k+1,1} - g_{k,1})\phi^p_{k+1}(n+2)\right|,$$

where

$$\phi^p_{k+1}(n+2) = \psi_1(n+2)\cdots\psi_k(n+2)f^p[x_{n+2},\ldots,x_{n+2-k}].$$

Since we are assuming that the derivative being approximated by the divided difference is nearly constant, so is the divided difference. In particular, the divided difference appearing in $\phi^p_{k+1}(n+2)$ can be approximated by that already computed in $\phi^p_{k+1}(n+1)$. The $\psi_i(n+2)$ and the $g_{k+1,1}, g_{k,1}$ depend only on the preceding step sizes, so we can compute them and predict the error. In exactly the same way we can predict the error that would be made on repeating a failed step.

What we really wish to do is to find the largest step size for which the predicted local error is acceptable. The only immediately apparent way to do this is by trial and error. This is quite unsatisfactory since the computation of the terms depending on the step size is rather expensive. In particular the $g_{k+1,1}$ and $g_{k,1}$ are expensive to compute.

Considering all this work and the skepticism the predictions deserve, it is reasonable to ask for a simpler, though perhaps less precise, way to compute these terms. If $h = h_{n+1}$ and we wish to consider $h_{n+2} = rh$, we can pretend that the preceding steps were all taken with a constant step size rh. If this were to be the case and we were to use the constant approximation to the divided difference, then we would have

$$\phi^p_{k+1}(n+2) = (rh)(2rh)\cdots(krh)f^p[x_{n+2},\ldots,x_{n+2-k}]$$
$$\doteq r^k\sigma_{k+1}(n+1)\phi^p_{k+1}(n+1),$$

on defining

$$\sigma_1(n+1) = 1,$$
$$\sigma_i(n+1) = \frac{h\cdot 2h\cdots(i-1)h}{\psi_1(n+1)\cdot\psi_2(n+1)\cdots\psi_{i-1}(n+1)} \qquad i = 2, 3,\ldots.$$

When the step size is constant, $g_{k+1,1} - g_{k,1} = \gamma_k^*$ so we predict the error using the step size $h_{n+2} = rh$ to be

$$r^{k+1}h\gamma_k^*\sigma_{k+1}(n+1)\phi_{k+1}^p(n+1).$$

This is a very convenient estimate. The quantities $\sigma_i(n+1)$ can be readily formed along with the other basic quantities of the implementation. If n_s steps have been taken with the constant step size h, we must have $\psi_j(n+1) = jh$ for $j \le n_s$ hence

$$\sigma_i(n+1) = 1 \qquad\qquad\qquad i \le n_s + 1,$$

$$= (i-1)\alpha_{i-1}(n+1)\sigma_{i-1}(n+1) \qquad i = n_s + 2, \dots.$$

In the code we form

$$\text{ERK} = \left|h\gamma_k^*\sigma_{k+1}(n+1)\phi_{k+1}^p(n+1)\right|.$$

It estimates what the local error at x_{n+1} would have been had the preceding steps been of constant size h. If the step is a success and we wish to consider a step size rh for the step to x_{n+2}, we predict that the local error will be r^{k+1} ERK in magnitude. If the step is a failure and we wish to repeat the step from x_n with a step of length rh, we also predict that this error will be r^{k+1} ERK in magnitude. It is very convenient to be able to use one technique in either case. Obviously, it is easy to consider whatever r we like because of the simple way it appears in the estimates.

The great convenience of this approach to prediction of errors is the main reason for its use in the codes of Krogh, Gear, and of this text. It is quite natural, since this is precisely what is done with the Taylor series methods that we have been relying upon for guidance. The step sizes are not allowed to vary rapidly and the terms which depend on the relative spacing of the mesh points are not sensitive to small changes in the spacing. Accordingly, we are interested only in values of r near 1, so that the pretended mesh spacing will not differ greatly from the actual one. Later we shall return to the question of how much these terms depend on the spacing. We shall design our algorithms so that they neither expect nor require very accurate predictions.

ORDER SELECTION

Since the order selection involves rather different ideas than the step size selection, they are done separately in the code. Besides ERK we also form

$$\text{ERKM1} = \left| h\gamma^*_{k-1}\sigma_k(n+1)\phi^p_k(n+1) \right|,$$

$$\text{ERKM2} = \left| h\gamma^*_{k-2}\sigma_{k-1}(n+1)\phi^p_{k-1}(n+1) \right|$$

which estimate what the local error at x_{n+1} would have been had the preceding steps been at the constant step size h and if the step to x_{n+1} had been taken at orders $k-1$ and $k-2$, respectively. Just as with ERK, they also predict the local error at x_{n+2} if we were to take a step of length h. To consider raising the order we need to predict what the error would be at order $k+1$. This is done only if the step to x_{n+1} is successful and only if we are actually using a constant step size h. The estimate requires that the step to x_{n+1} be completed since it uses $\phi_{k+2}(n+1)$. In view of the analysis of the preceding chapter these additional conditions seem prudent. Just as before, the predicted error at x_{n+2} with a step size of h is

$$\text{ERKP1} = \left| h\gamma^*_{k+1}\phi_{k+2}(n+1) \right|.$$

Since we are pretending to have a constant step size h, the numerical solution is

$$y_{n+2} = y_{n+1} + hf^p_{n+2} + h\gamma^*_1\nabla f^p_{n+2} + h\gamma^*_2\nabla^2 f^p_{n+2} + \cdots.$$

This is analogous to a Taylor series for the local solution and we estimate the error of stopping with any given term by the term that follows it in the series. The basic idea of selecting an order is to use terms until they start to increase. The size of the terms depends on the step size that is used and we have not yet chosen the step size. However, we wish to approximate the local solution uniformly well over the entire interval $[x_{n+1}, x_{n+2}]$ so the decision should not depend very much on the step size h_{n+2}. It seems reasonable to base the order choice on an "average" step size and more on the qualitative properties of the terms than on their actual sizes. The step size h_{n+2} will be permitted to vary only between $\frac{1}{2}h$ and $2h$, so h is a reasonable choice to represent the interval. The philosophy of the order selection is to change the order only if the predicted error is reduced and if there is a trend in the terms. At each step we consider lowering the order from k to $k-1$. To gain an immediate advantage we must have ERKM1 < ERK. A trend is signaled by ERKM2 < ERK. For several reasons we prefer to use the lowest order which appears satisfactory so we shall also lower the order if there is just equality. If $k=2$, we do not have ERKM2 available so we require a stronger indication of an advantage before lowering the order. Thus we lower the order if

$$k > 2 \quad \text{and} \quad \max(\text{ERKM1}, \text{ERKM2}) \leq \text{ERK},$$

$$k = 2 \quad \text{and} \quad \text{ERKM1} \leq 0.5\,\text{ERK}.$$

If we have an estimate of the error at order $k + 1$ available, we ask if there is an immediate advantage, $\text{ERKM1} \leq \text{ERK}$, and a trend, $\text{ERKM1} \leq \text{ERKP1}$. Thus we lower the order if

$$\text{ERKM1} \leq \min(\text{ERK}, \text{ERKP1}).$$

The estimates ERKM2, ERKM1, ERK are formed at the same time we form ERR, so lowering the order is considered whether or not the step is successful. If the step is unsuccessful, the final evaluation is never done and the cost of a failure is reduced. ERKP1 uses the final evaluation, and so it is formed only on successful steps when a constant step size is used. Because our philosophy calls for an advantage before changing the order, we do not consider raising the order after failed steps. Not only do we wish to avoid a final evaluation which would be discarded after forming ERKP1, but it also seems unreasonable to add data points when a failure indicates a lack of smoothness of the solution.

The order is raised only if an advantage is indicated by $\text{ERKP1} < \text{ERK}$ and a trend by either $\text{ERKP1} < \text{ERKM1}$ or $\text{ERKP1} < \text{ERKM2}$. Because of the test for lowering the order, the coding for this is to raise the order if

$$1 < k < 12 \quad \text{and} \quad \text{ERKP1} < \text{ERK}.$$

It is done this way because if it is possible to lower the order, we do so. To reach the test for raising the order it must be the case that

$$\text{ERK} < \max(\text{ERKM1}, \text{ERKM2}),$$

hence

$$\text{ERKP1} < \text{ERK} < \max(\text{ERKM1}, \text{ERKM2}).$$

When $k = 1$ we ask for a stronger indication of an advantage. We raise the order if

$$k = 1 \quad \text{and} \quad \text{ERKP1} < 0.5\,\text{ERK}.$$

With two exceptions, this specifies the order selection algorithm. The starting phase of the computation is a special case which we discuss after the step size selection. The second set of exceptions which call for a drastic reduction in the order is detected by repeated step failures, so we shall discuss them after going into the step size algorithm.

The conversion of the error estimates to constant step size is not important to the order selection algorithm. Often the code is already using a constant step size and there is no conversion. Since it is the qualitative behavior of the terms which affects the decisions about the order, the distribution of the mesh points is not critical. The "best" order has not been precisely defined and in fact is not critical to the computation. Frequent changes in the order can occur only when the step size is constant, since this is a condition for raising the order. In this situation small reductions in the estimated error may cause the order to fluctuate, which in turn helps the code continue with constant step size. In any situation a change in order is quite inexpensive.

There are several important situations recognized and properly handled by the order adjustment. The validity of the estimates requires that the corresponding derivatives be slowly varying. If one of these is not, then a higher order derivative must be relatively large, as must a corresponding difference. An extreme instance is a discontinuity in a derivative. Behavior of this kind will cause a reduction of the order until the corresponding derivatives of the function are reasonably smooth. A large jump in a low order derivative cannot be handled so simply and there is special provision in the step size algorithm for this. There is also a kind of practical instability which is studied in Chapter 8. One way it manifests itself is that low order formulas are as accurate as high order formulas. The algorithm will therefore tend to use low orders, which is fortunate since they have the better stability properties. We shall go into the matter further in Chapter 8. The machine arithmetic causes errors in the basic function evaluations which propagate into the differences. They tend to make the differences increase as a function of their order, as we shall argue in Chapter 9. As a result the order selection rules tend to use only those differences which still have some meaningful figures.

STEP SIZE SELECTION

After selecting the order to be used, we rename it k, and the corresponding estimated error at constant step size, ERK. The predicted error at x_{n+2} after a step of size rh is then r^{k+1} ERK. A locally optimal choice of r is to take it as large as possible while still satisfying r^{k+1} ERK $\leq \varepsilon$. This leads to

$$r = \left(\frac{\varepsilon}{\text{ERK}}\right)^{\frac{1}{k+1}}.$$

One cannot accept this "optimal" step size without reservation. The basic approach is a reasonable one, and works well in practice, but the estimated r can be crude. We have emphasized the possible unreliability of the conversion of the local error estimates to constant step size. Values of r which are either large or small cannot really be justified by the arguments employed. We shall attempt to nullify potentially unreliable estimates by our algorithm. This protection generally leads to a smaller step size than is necessary when the estimates are good. Still, inefficiency due to a conservative step size tends to be redeemed by fewer step failures and smaller global errors. Even if the value of r is reliable, we must impose some limits on the change of step size. For one thing we must limit r (for orders greater than one) to stabilize the integration in the practical sense discussed in Chapter 4. For another, the optimal step size is being estimated from the past behavior of the solution and may not be appropriate for the future, e.g. a large increase of the step size may appear justified because of a derivative passing through zero but lead to an immediate step failure.

Choosing r so that the predicted error is equal to the tolerance ε is certain to lead to many instances of the error being too large. To prevent failures of this sort we are conservative and choose an r which predicts an error of 0.5ε:

$$r = \left(\frac{0.5\varepsilon}{\text{ERK}}\right)^{\frac{1}{k+1}}. \tag{1}$$

For reasons already cited we limit increases to a factor of two. Thus in the code we first see if it is possible to double the step size, that is, if $r \geq 2$. It is expedient to do this by asking if

$$0.5\varepsilon \geq 2^{k+1}\,\text{ERK}.$$

Such a test avoids the awkwardness of an ERK which vanishes and, by referring to a table of 2^{k+1}, is cheaper than computing r. The upper limit of 2 is a compromise between reasons for limiting the increase and the desire to adapt to the solution as fast as possible. After the start phase, one expects the appropriate step size to change gradually. However, because an adjustment of the step size is considered at every step, quite a rapid increase of the step size is possible if it really appears justified.

If it is not possible to double the step size, we do not increase it at all. By foregoing increases of a factor smaller than 2, we strongly bias the algorithm towards a constant step size. This action has many important consequences. It provides a better theoretical foundation for the error estimates and predictions, reduces the overhead, allows the possibility of increasing the order to be considered often enough, and is also quite conservative.

To see if it is possible to retain the current step size, we ask if

$$0.5\varepsilon \geq \text{ERK}.$$

If not, we must reduce the step size and we do so by the factor r obtained from (1), with some qualifications. The factor is not permitted to be greater than 0.9. This conservative approach assures us that if we must reduce the step size, we do so by a non-trivial amount. The factor is not permitted to be smaller than 0.5. Along with the restriction on step size increases this guarantees the stability of the algorithm with respect to the changing of step size. The lower bound is ordinarily superfluous since the estimates seldom result in a smaller factor. (The step being successful, we must have $\varepsilon \geq \text{ERR}$. If the order was not changed, then $\text{ERK} \doteq \text{ERR}$ or else the value of r is unreliable and should be ignored anyway. If the order was changed, then because of the renaming and the order selection algorithm, ERK is less than a quantity approximately equal to ERR. In any event we normally expect that ERK is less than or roughly equal to ε. Halving the step size leads to a predicted error of $(\frac{1}{2})^{k+1}$ ERK which we can therefore expect to be less than $\frac{1}{2}\varepsilon$.) Of course, if there is a sudden change in the character of the solution, a larger reduction of the step size may be appropriate. We cannot really expect to predict such a change and we cope with it after a failed step.

The strategy for reducing the step size is rather different from that for increasing it. In reducing it we appear to take the estimate for r at face value and in increasing it we scarcely presume it to be accurate at all. We are not more sanguine about the optimal step size when reducing the step size, our intention is just a little different. Estimating an optimal step size smooths out the step sizes being used and allows the code to arrive at a "best" value in a way that is impossible with the use of halvings and doublings, only.

We are even less confident of the error estimates when the step is a failure, since they use an unacceptable solution value. Our approach is simply to halve the step size and try again. In view of the step size and order adjustments made on the preceding steps a failed step is ordinarily due to a somewhat abrupt change in the solution. Repeated failures to complete a step successfully are a clear signal that the solution is discontinuous.

If some very low order derivative of the solution has a large jump, the problem does not fall in the basic class treated in this text. The extension of the theory sketched in Chapter 1 says that if we restart at the point of discontinuity we then have two problems, each of which is appropriate for our methods. The practical question is whether a code can locate such a discontinuity automatically: it is better if the user does this and restarts the computation himself but the code provided with this text does quite well on its own. The order selection algorithm described so far lowers the order by at most one after an unsuccessful attempt at a step. We add the rule that after three tries the order is dropped to one. This in effect restarts the code. Since the method of order one has no memory, any step taken at this order can be regarded as a fresh starting point. For stringent tolerances the code must find the discontinuity in order to pass it and start increasing the step size and order again. To facilitate this the code starts choosing either a half step or the optimal step size, whichever is smaller. There is virtually no overhead associated with order one and since it is essentially the Taylor series method, the estimate is usually reliable.

THE START PHASE

In each integration there is an initial phase during which a step size and order appropriate to the problem are found. Our code begins with the Adams formula of order one, requiring no memorized values, and increases the order as sufficient data become available. Most codes require that the user guess an initial step size but this is rather unsatisfactory. A first order method needs a step size that seems appallingly small to most users and, besides this, most users have no idea of how to estimate a suitable value. Since there is an adequate way to estimate this value automatically, there is no need to trouble the user, and our codes do not.

Suppose we started with a formula of order zero, that is $y_1 = y_0 = A$. Then

$$y(x_1) = y(a) + hy'(a) + O(h^2)$$
$$= A + hf(a, A) + O(h^2),$$

and here $u_0(x) = y(x)$, so the local error

$$u_0(x_1) - A = hf(a, A) + O(h^2).$$

As a rule of thumb, let us suppose that the error of the first order method is h times the error of the zero order method. Then, since we want

$$0.5\varepsilon \doteq |\text{le}_1| \doteq h^2 |f(a, A)|,$$

we should take

$$h \doteq \left| \frac{0.5\varepsilon}{f(a, A)} \right|^{\frac{1}{2}}.$$

It is desirable to be conservative so we use one quarter of this value. The user of the code STEP must provide an upper bound, h_{input}, on the size of the step. This bound indicates the general scale of the problem and prevents any difficulty due to $y'(a)$ being zero. Thus we use

$$h = \min\left(h_{\text{input}}, \frac{1}{4} \left| \frac{0.5\varepsilon}{f(a, A)} \right|^{\frac{1}{2}} \right).$$

This rule of thumb for the initial step size works adequately, in part because the code is insensitive to its value. A step size conservative for a method of order one is chosen, but because of extrapolation the approximate solution comes from a method of order two. This means that the smaller the tolerance ε is, the more conservative the initial step size is likely to be. We shall take advantage of this fact in designing an algorithm for choosing the step size and order during the start.

The code will perform well during the first phase of the integration without the introduction of additional rules for the choice of step size and order. The typical behavior is that, after reaching a given order k, the step size will double on successive steps until a step size appropriate to the order is reached. Then $k + 1$ steps of this size are taken and the order is raised. This behavior is repeated until an order appropriate to the problem and tolerance is found. There is some inefficiency due to the caution in raising the order, so some additional rules are introduced. The first phase of the integration is indicated in the code STEP by a logical variable PHASE1 set .TRUE.. During this phase we double the step size and raise the order after every step. We leave this first phase of the computation whenever there is a step failure, or the order selection algorithm says to lower the order, or the maximum order of 12 is reached.

The principal reason for these rules is to increase the order more boldly than for typical steps. Their success depends on prior knowledge that the order should be increased quickly and on the fact that lowering the order is considered at every step. The increased efficiency is important only when high orders are appropriate, which ordinarily corresponds to small tolerances ε. Since for small ε the initial step size is especially conservative, doubling the step size at each step still leaves the step size much smaller than could be used at the higher orders. As a result, the smaller ε is, the more likely it is that the code will leave the start phase because the proper order is found, rather than because of a step failure. This is relevant to the code's efficiency because the algorithm for typical steps is used after a step failure. In particular, if the initial step size is too big, the new rules do not come into play at all. The typical behavior is to leave the start phase because of a signal to lower the order and then to double the step size for several successive steps to reach a value appropriate to the final order.

The reader might think it would be a good idea to restart with these additional rules after the code detects a discontinuity. It would be except that the code may detect stiffness in the same way and, as we shall see, a rapid increase of the order is then disastrous. Because both of these difficulties are rare, the usual rules perform quite satisfactorily and additional complication of the code is unwarranted.

IMPLICATIONS OF THE ALGORITHMS

One consequence of the step size algorithm is a strong bias towards constant step size. The discretization error associated with $y_{n+1}(k)$ is estimated for constant step size h. From Chapter 6 this must be approximately the local truncation error of the corrector so the step size *roughly* satisfies the relation

$$0.5\varepsilon \doteq \left| h_{n+1}^{k+1} \gamma_k^* y^{(k+1)}(x_n) \right|. \tag{2}$$

Since $y^{(k+1)}$ is slowly varying, the "optimal" step size is also changing slowly for "small" step sizes. Because the step size will not be increased until it can be doubled, we are assured that there will be long stretches of constant step size, at least for "small" step sizes. The fact that the step size is reduced by at least 10 percent also assists in bringing about this behavior.

Limiting changes of step size to halvings and doublings implies that a limit is placed on the mesh distortion. In turn, this means that it is not unreasonable to predict the discretization error by pretending that a constant step size is being used. The discretization error of $y_{n+1}(k)$ is, from Chapter 3,

$$\frac{u_n^{(k+1)}(\xi)}{k!} \int_{x_n}^{x_{n+1}} (t - x_{n+1})(t - x_n)\cdots(t - x_{n+2-k}) \, dt.$$

If $h = h_{n+1}$, then the limits on the increase of step size imply

$$h_n \geq \tfrac{1}{2}h,$$

$$h_{n-1} \geq \tfrac{1}{2}h_n \geq \frac{h}{2^2},$$

$$\vdots$$

$$h_{n+2-k} \geq \frac{h}{2^{k-1}}.$$

If we grossly underestimate the terms in the integral by assuming that the smallest step size is used at each step, we have

$$-\frac{1}{k!} \int_{x_n}^{x_{n+1}} (t - x_{n+1})(t - x_n) \cdots (t - x_{n+2-k}) \, dt \geq -\frac{h^{k+1}}{2^{k-1}} \gamma_k^*.$$

Using the same argument with the largest possible step size leads to a bound of

$$-2^{k-1}h^{k+1}\gamma_k^*.$$

It is clear from their derivation that these bounds are very crude, especially so in view of the infrequency of step size changes. From them we see that the estimate based on constant step size, $h^{k+1}\gamma_k^* u_n^{(k+1)}(\xi)$, is never terribly far from the value obtained by using the true step sizes.

Let us consider a sequence of steps of constant step size and order. The step size at order k roughly satisfies equation (2) so we should have

$$\varepsilon \geq \left| h_{n+1}^{k+1} \gamma_k^* y^{(k+1)}(x_n) \right|.$$

Arguing just as in the estimation of $le_{n+1}(k+1)$ we find that the local error actually made at each step consists mainly of discretization (truncation) error. Therefore,

$$\left| \text{LE}_{n+1} \right| \doteq \left| h_{n+1}^{k+2} \psi_k(x_n) \right| = h_{n+1} \left| \psi_k(x_n) \right| h_{n+1}^{k+1} \leq h_{n+1} \varepsilon \frac{\left| \psi_k(x_n) \right|}{\left| \gamma_k^* y^{(k+1)}(x_n) \right|},$$

where $\psi_k(x)$ is the local truncation error of the PECE procedure of order k. From this we see it is approximately true that

$$\left| \text{LE}_{n+1} \right| \leq \sigma \varepsilon h_{n+1},$$

for a suitable constant σ, and that the control is a generalized error per unit step.

GLOBAL ERROR ESTIMATION

As was discussed in the preceding chapter, one way to view the numerical approximation $y_I(x)$ is as the exact solution of an equation close to the original one,

$$y_I'(x) = f(x, y_I(x)) + r(x)$$

or

$$y_I'(x) = (1 + \rho(x))f(x, y_I(x)) + \tau(x),$$

where $\rho(x)$ is a relative error and $\tau(x)$ an absolute error. Let us consider how to measure and report the size of the residual. A convenient way is to find a number μ such that

$$|\rho(x)| \le \mu \quad \text{and} \quad |\tau(x))| \le \mu$$

for this represents a mixed relative-absolute measure much like that used in one of our codes for the control of local error. The smallest μ for which these bounds are true is

$$\mu = \max_{[a,b]} \frac{|r(x)|}{1 + |f(x, y_I(x))|}.$$

We can determine the value of $r(x)$ anywhere but it costs a function evaluation to do so. We must compromise between a sharp estimate of μ and a reasonable cost. At each mesh point

$$r(x_n) = y_I'(x_n) - f(x_n, y_I(x_n)) = f_n - f(x_n, y_n) = 0,$$

so it seems adequate to evaluate $r(x)$ at the midpoint of each successful step. For $x_{\text{mid}} = (x_n + x_{n+1})/2$ we interpolate to find $y_I(x_{\text{mid}})$, $y_I'(x_{\text{mid}})$, evaluate $f(x_{\text{mid}}, y_I(x_{\text{mid}}))$, and finally we evaluate $r(x_{\text{mid}})$. Thus we take

$$\mu \doteq \max \frac{|r(x_{\text{mid}})|}{1 + |f(x_{\text{mid}}, y_I(x_{\text{mid}}))|}$$

as we step from a to b.

This procedure uses one function evaluation per step and so it increases the cost over just computing the solution by about 50%. In our experience, a good Adams code, when applied to physical problems, always leads to a value of μ about the size of the user's error tolerance, so it is hard to justify the expense of verifying this routinely.

A way to assess global errors is to reintegrate the problem with a reduced tolerance and to compare the results. Suppose y_{ε_1} is the approximation to $y(b)$ obtained using local error tolerance ε_1, and y_{ε_2} that using $\varepsilon_2 < \varepsilon_1$. Let

$$e_1 = y(b) - y_{\varepsilon_1},$$

$$e_2 = y(b) - y_{\varepsilon_2},$$

$$est = y_{\varepsilon_2} - y_{\varepsilon_1}.$$

The idea is to estimate e_1 by est. Now

$$e_1 - e_2 = est$$

so

$$\left| \frac{est}{e_1} \right| = \left| 1 - \frac{e_2}{e_1} \right|.$$

Suppose that the code is effective in the sense that reducing the tolerance actually reduces the global error. We must take ε_2 enough smaller than ε_1 that $|e_2|$ is substantially smaller than $|e_1|$; but of course the smaller ε_2, the more the second integration costs. A good practical value is $\varepsilon_2 = 0.1\varepsilon_1$. In a variable order code like ours the cost increases rather slowly as the tolerance is reduced, and with this choice, for typical problems, the additional integration costs about one and one-fifth times the original one. Supposing that $|e_2| \le \beta |e_1|$ for some $\beta < 1$ we then find that

$$1 - \beta \le \left| \frac{est}{e_1} \right| \le 1 + \beta.$$

For example if $\beta = 0.5$, the estimate is off by no more than a factor of two and if $\beta = 0.9$, it is off by no more than an order of magnitude.

There are many numerical results presented in Chapter 11 for our codes. They suggest that if $\varepsilon_2 = 0.1\varepsilon_1$, then it is usually true that $|e_2| \le \beta |e_1|$ with $\beta = 0.5$ and that with rare exceptions one can take $\beta = 0.9$. Thus this approach leads to a fairly reliable estimate of the global error to within a factor of two and a reliable estimate to within an order of magnitude. In fact there is good reason to expect that β ought to be roughly 0.1. The bound on the global error in Chapter 6 is proportional to the tolerance ε. Using 0.1ε reduces the bound by the same factor. If one argues along these lines with estimates rather than bounds, it will be found that the global error itself (rather than the bound) is proportional to ε. We shall only sketch the ideas because the approximations are rough and we make no serious use of the conclusion.

The global error e_{n+1} can be split into

$$e_{n+1} = y(x_{n+1}) - u_n(x_{n+1}) + \text{LE}_{n+1}.$$

From the equation of first variation

$$u_n(x_{n+1}) \doteq y(x_{n+1}) - v(x_{n+1})(y(x_n) - y_n)$$

where

$$v' = f_y(x, y(x))v, \qquad v(x_n) = 1.$$

Letting $g(x) = f_y(x, y(x))$ we solve for $v(x)$ and arrive at

$$y(x_{n+1}) - u_n(x_{n+1}) \doteq e_n \exp\left(\int_{x_n}^{x_{n+1}} g(t)\, dt\right)$$

and then

$$e_{n+1} \doteq e_n \exp\left(\int_{x_n}^{x_{n+1}} g(t)\, dt\right) + \text{LE}_{n+1}.$$

Repetition of this approximate equality shows that

$$e_n \doteq \sum_{j=1}^{n} \text{LE}_j \exp\left(\int_{x_j}^{x_n} g(t)\, dt\right).$$

The step size h_j is determined so that

$$0.5\varepsilon \doteq h_j^{k+1} \left|\gamma_k^* y^{(k+1)}(x_j)\right|$$

when at order k. Similarly

$$\text{LE}_j \doteq h_j^{k+2} \psi_k(x_j)$$

where $\psi_k(x)$ is the local truncation error of the PECE procedure of order k. Then

$$\text{LE}_j \doteq h_j \psi_k(x_j) h_j^{k+1} \doteq \frac{0.5\varepsilon h_j \psi_k(x_j)}{\left|\gamma_k^* y^{(k+1)}(x_j)\right|}$$

and

$$e_n \doteq 0.5\varepsilon \sum_{j=1}^{n} h_j \left[\frac{\psi_k(x_j)}{|\gamma_k^* y^{(k+1)}(x_j)|} \exp\left(\int_{x_j}^{x_n} g(t)\, dt \right) \right].$$

This sum can be regarded as a Riemann sum for small h_j (small ε) which is approximately evaluated as an integral:

$$e_n \doteq 0.5\varepsilon \int_{a}^{x_n} \frac{\psi_k(s)}{|\gamma_k^* y^{(k+1)}(s)|} \exp\left(\int_{s}^{x_n} g(t)\, dt \right) ds.$$

This approximate expression for the global error assumes that the order is held constant. To account for a variation of the order we can suppose that the basic interval is split into subintervals on which the order is constant and then apply this approximate analysis to each. The point is that, in this approximation, the global error is proportional to the local error tolerance ε. The conclusion helps support our contention that the method of reintegration furnishes a reliable, though crude, estimate of global errors when using the codes provided.

8

STABILITY—
LARGE STEP SIZES

The typical solution of an initial value problem involves only one or, at most, a few integrations. The analysis of Chapter 4 considers these integrations as embedded in a limit process where the maximum step size tends to zero. It is shown that for "sufficiently" small step sizes the global errors are small. There is no reason to expect small errors or even qualitative agreement between the approximate and true solutions when the step size is too large. For some problems the step size is restricted by the requirement that the numerical solution behave qualitatively like the true solution. If the step size is not so restricted, the computation is said to be "unstable." In both Chapter 4 and this chapter we are concerned with restrictions on the step size that make the numerical approximations reasonable. But the circumstances are different—the limit of a sequence of integrations is considered in Chapter 4, while in this chapter what happens in a single integration is considered.

Codes like those discussed in this text rarely exhibit the catastrophic instability which is possible. Although it is necessary to deal with special cases to be precise, the general idea is plausible and accounts for the behavior observed. Instability is simply a growth of error, which is a result of the step size being so large that the formulas do not give reasonable approximations. The error estimators have been proved valid only when the step size is small.

It is plausible, and usually true, that even for large step sizes, the error estimators will detect large errors and force a reduction in the step size. If the errors can be detected in this way, the trouble will be cured by reducing the step size to the point at which reasonable approximations are again obtained. The efficiency of such a code will be limited by the stability properties of the formulas employed; but generally speaking, the accuracy of reported results is not affected.

Normally the accuracy requirement keeps step sizes in a range where instability does not occur but there are problems for which stability limits the step size. In extreme cases, the so-called stiff problems, a code based on Adams methods is inefficient for just this reason, and it becomes more appropriate to use other methods with better stability properties. Roughly speaking, a stiff problem is one in which the solution components of interest are slowly varying but solutions with very rapidly changing components are possible. An example is an attitude control system for a rocket designed to maintain a smooth flight path which will very rapidly correct any deviation from this path. Another example is a mixture of chemically reacting species in which reactions are taking place slowly but in which some species have the capacity to react quickly. Methods and a code aimed specifically at stiff problems can be found in [3]. Although stiff problems are physically important, most practical problems are not stiff. Their solution by methods aimed specifically at non-stiff problems is so much simpler and more effective, that codes for both purposes are necessary at this time.

Codes of the kind studied in this text also vary their order to reduce local errors while using an efficiently large step. The order selection algorithm is biased towards using low orders if they are sufficiently accurate. This bias causes the codes to select low orders when a problem becomes stiff, since the accuracy requirement is then easy to satisfy. When the analysis is feasible we shall see that the lower the order, the more stable the method, at least for the Adams methods. With this as a working hypothesis and with a general understanding of the interaction of instability and the local error estimator, we can predict what the behavior of these codes will be when they encounter difficulty with stability: the code will tend to reduce its order and step size, especially the order, to cope with the difficulty. Stiff problems will drive the order and the step size down to small values. Extreme difficulty can cause the code to exit on minimum step size but more frequently the computation is terminated because it has become too expensive. Rarely, if ever, will instability be evidenced by undue error growth. These predictions are in agreement with computational experience and we shall use them to devise an indicator of stiffness in the driver DE. This will be discussed further after we investigate how such codes detect instability.

To gain insight into the meaning of stability and to compare the stability properties of various methods we shall have to consider simple problems for which the analysis is feasible. The conclusions are rigorously valid only for the models employed but they are suggestive of the behavior in the general case. The model problems to be treated are of the form

$$\mathbf{y}' = A\mathbf{y} + \mathbf{g}(x) \tag{1}$$

for a constant matrix A. A solution $\mathbf{y}(x)$ of the general nonlinear problem $\mathbf{y}' = \mathbf{f}(x, \mathbf{y})$ can be approximated near a point x_0 by

$$\mathbf{y}' \doteq f_y(x_0, \mathbf{y}(x_0))(\mathbf{y} - \mathbf{y}(x_0)) + \mathbf{f}(x_0, \mathbf{y}(x_0)) + \mathbf{f}_x(x_0, \mathbf{y}(x_0))(x - x_0)$$

where f_y is the Jacobian matrix

$$f_y = \left(\frac{\partial \mathbf{f}}{\partial y_1}, \frac{\partial \mathbf{f}}{\partial y_2}, \ldots, \frac{\partial \mathbf{f}}{\partial y_n} \right)$$

and $\mathbf{f}_x$ the vector

$$\left(\frac{\partial f_1}{\partial x}, \frac{\partial f_2}{\partial x}, \ldots, \frac{\partial f_n}{\partial x} \right)^T .$$

This suggests that for a few steps near x_0 the behavior of a nonlinear problem may be simulated by a model problem of the form (1) but we shall put the matter no more strongly than this. We prefer to think of the model problem as one on which codes ought to perform adequately and which possibly illuminates the more general situation.

A discussion of stability must be based on an analysis of how error propagates during the course of a single integration. It is suggestive to consider first how isolated errors in following the solution curve are propagated by the differential equation itself. If at some point x_r the initial value of the differential equation (1) is changed from $\mathbf{y}(x_r)$ to $\mathbf{y}(x_r) + \boldsymbol{\delta}(x_r)$, the solution is changed to $\mathbf{u}(x) = \mathbf{y}(x) + \boldsymbol{\delta}(x)$. By substitution, we see that $\boldsymbol{\delta}(x)$ is the solution of the problem

$$\boldsymbol{\delta}(x_r) \text{ given,}$$

$$\boldsymbol{\delta}' = A\boldsymbol{\delta}, \qquad x \geq x_r. \tag{2}$$

Furthermore, if there is an additional error of $\boldsymbol{\mu}(x_j)$ at $x_j > x_r$, the solution is changed to $v(x) = y(x) + \delta(x) + \mu(x)$ for $x \geq x_j$ where $\mu(x)$ is the solution of

$$\mu(x_j) \text{ given,}$$

$$\mu' = A\mu, \qquad x \geq x_j.$$

This tells us that the effect of independent errors is just additive, so for analytical purposes it suffices to consider the effect of a single error. These observations suggest that when using the extrapolated Adams formulas of order k,

$$p_{n+1} = y_n + h \sum_{j=1}^{k} \alpha_{k,j} f_{n+1-j},$$

$$y_{n+1} = y_n + h \sum_{j=1}^{k} \alpha^*_{k+1,j} f_{n+1-j} + h\alpha^*_{k+1,0} f^p_{n+1},$$

a block of k independent successive errors $\delta_r, \delta_{r+1}, \ldots, \delta_{r+k-1}$ propagate as the numerical solution of (2). Furthermore the effect of independent errors is additive, so the consideration of a single block of initial errors can show what will happen in general. We shall show the first property to be true and leave the proof of the second to the reader. Given the initial values

$$y_r + \delta_r, y_{r+1} + \delta_{r+1}, \ldots, y_{r+k-1} + \delta_{r+k-1}$$

the formulas applied to (1) are

$$(p_{n+1} + \delta^p_{n+1}) = (y_n + \delta_n) + h \sum_{j=1}^{k} \alpha_{k,j} [A(y_{n+1-j} + \delta_{n+1-j}) + g_{n+1-j}],$$

$$(y_{n+1} + \delta_{n+1}) = (y_n + \delta_n) + h \sum_{j=1}^{k} \alpha^*_{k+1,j} [A(y_{n+1-j} + \delta_{n+1-j}) + g_{n+1-j}]$$

$$+ h\alpha^*_{k+1,0} [A(p_{n+1} + \delta^p_{n+1}) + g_{n+1}]$$

in clear notation. Using the definition of the y_n, we see that the δ_n are the solution of the problem

$$\delta_r, \delta_{r+1}, \ldots, \delta_{r+k-1} \quad \text{given,}$$

$$\delta^p_{n+1} = \delta_n + h \sum_{j=1}^{k} \alpha_{k,j} A\delta_{n+1-j}, \tag{3}$$

$$\delta_{n+1} = \delta_n + h \sum_{j=1}^{k} \alpha^*_{k+1,j} A\delta_{n+1-j} + h\alpha^*_{k+1,0} A\delta^p_{n+1}.$$

This is just the numerical solution of (2) as claimed.

Throughout the text we have worked with a single equation rather than a system of equations because there is no essential difference and the details are simpler. This reasoning is still valid, but an additional element enters. The theory of matrices asserts that for a given matrix A, it is nearly always the case that there exists a matrix T such that $TAT^{-1} = D$, where $D = \text{diag } \{d_1, d_2, \ldots, d_n\}$ is a diagonal matrix with the eigenvalues of A as the diagonal elements. Matrices A which cannot be diagonalized in this way can be represented as the limits of matrices which can. Because initial value problems for (1) are well posed we can change A by an arbitrarily small amount to get a matrix which can be diagonalized and, in doing this, change the solutions by an arbitrarily small amount. Here and later we shall ignore those special limit cases which complicate the analysis but do not affect the conclusions. The matrix T represents a change of variables which splits the system into independent, single equations. For, if we let

$$\mathbf{z}(x) = T\mathbf{y}(x), \quad \mathbf{G}(x) = T\mathbf{g}(x),$$

then

$$\mathbf{z}' = T\mathbf{y}' = TAT^{-1}T\mathbf{y} + T\mathbf{g}(x) = D\mathbf{z} + \mathbf{G}(x). \tag{4}$$

In component form (4) becomes

$$z_i'(x) = d_i z_i(x) + G_i(x) \qquad i = 1, 2, \ldots, n. \tag{5}$$

It is a key fact that the same transformation affects the numerical methods the same way; in particular, it can be easily verified that if we let $\Delta_n = T\delta_n$, the error propagated in the solution of the transformed equation (4) is just the transformed version of (3). The point is that the solution of a single equation

$$y' = Ay + g(x) \tag{6}$$

might as well be examined, rather than the general form (1), provided that A is permitted to be a complex number, since eigenvalues in general are complex.

Whether or not a computation is stable depends on the error criterion and the solution, as well as on the behavior of the propagated error. To see this, consider the equation $y' = -y$, $y(0) = 1$. If a pure absolute error criterion is used, then because the solution $y(x) = \exp(-x)$ tends to zero, eventually any number less than the error tolerance is an acceptable answer. The accuracy requirement has essentially fallen away and the step size is limited only by the stability requirement that the numerical solution, and accordingly its error, not grow too much. Because any number less than the tolerance is acceptable, the error is allowed to grow provided the numerical solution does not exceed the tolerance. Only when the solution exceeds the tolerance is the computation regarded as being unstable and requiring action. Quite a different view is obtained if it is pure relative error that concerns us. Because of the behavior of the solution, the absolute error must decrease in a sufficiently rapid manner that the relative error meets the tolerance. In this case a failure of the error to decrease fast enough can constitute instability.

Instability becomes a consideration only when the accuracy requirement does not restrict the step size sufficiently. When a solution is changing fairly rapidly and the error criterion forces a code to follow the curve, it appears that the accuracy requirement almost always dominates. The most important situation that has been singled out for analysis is that in which the equation is quite stable and all solutions ultimately become slowly varying. The general solution of (6) is

$$y(x) = c \exp(Ax) + \int_a^x g(t) \exp(A(x - t)) \, dt,$$

where the constant c is determined by the initial condition at $x = a$. If the real part of A is negative, all solutions converge on the particular solution

$$p(x) = \int_a^x g(t) \exp(A(x - t)) \, dt,$$

and so the differential equation is stable, the more so as $|A|$ is large. Assuming that $g(x)$ is such that the limit solution $p(x)$ is slowly varying, the qualitative behavior is clear from the explicit general solution and the expression

$$y^{(k+1)}(x) = A^{k+1} c \exp(Ax) + p^{(k+1)}(x)$$

for derivatives. After an initial period of rapid change all solutions tend to a particular one and become slowly varying. The extreme case of constant g is representative. The general solution is

$$y(x) = c \exp(Ax) - \frac{g}{A},$$

which tends to the constant particular solution

$$-\frac{g}{A}.$$

Furthermore

$$y^{(k+1)}(x) = A^{k+1}c\exp(Ax)$$

tends to zero exponentially fast as x increases. Since this derivative decreases rapidly for large x, the truncation errors of the Adams methods also decrease rapidly and accuracy ceases to limit the step size. Because $y(x)$ is essentially constant, instability measured by either absolute or relative error corresponds to the δ_n increasing in magnitude.

It may seem odd that one would continue integrating a problem of this sort after the solution becomes slowly varying. It must be kept in mind that the equation may be only one of a system and the integration is continued because it is another component that is of interest to us. With the system split into independent equations, we can see that the step size limitation is being imposed by an equation for which the accuracy requirement is no longer relevant. Since in practice we do not actually uncouple the equations in this fashion, and the whole analysis is only approximate for nonlinear systems anyway, we have no choice but to continue the integration.

The error estimators used are of the form

$$\sum \eta_i f_{n+1-i},$$

which for the model problem is

$$\sum \eta_i[A(y_{n+1-i} + \delta_{n+1-i}) + g_{n+1-i}]$$
$$= \sum \eta_i[Ay_{n+1-i} + g_{n+1-i}] + \sum \eta_i A\delta_{n+1-i}.$$

The error estimator sees the δ_n as the error arising from the numerical solution of (2). Because $Re(A) < 0$, the true solution $\delta(x)$ of (2) tends to zero, so any growth of the δ_n causes the error estimator to ultimately force a reduction of the step size.

To proceed further with the analysis, we make the usual assumption that the step size is a constant h since we can then obtain the δ_n sufficiently explicitly to determine whether or not instability is present. If we let $\lambda = hA$, the equations of (3) satisfied by the errors δ_n combine to give

$$\delta_{n+1} = \delta_n + \lambda \sum_{i=1}^{k} \alpha_{k+1,i}^* \delta_{n+1-i} + \lambda\alpha_{k+1,0}^* \left[\delta_n + \lambda \sum_{i=1}^{k} \alpha_{k,i}\delta_{n+1-i}\right]. \quad (7)$$

This is an example of a difference equation of order k with constant coefficients. A standard method of solution is to attempt a solution of the form

$$\delta_n = \zeta^n \qquad \text{all } n$$

for a constant ζ. Substitution shows that equation (7) is satisfied if

$$\zeta^{n+1} = \zeta^n + \lambda \sum_{i=1}^{k} \alpha^*_{k+1,i} \zeta^{n+1-i} + \lambda\alpha^*_{k+1,0}\left[\zeta^n + \lambda \sum_{i=1}^{k} \alpha_{k,i}\zeta^{n+1-i}\right]$$

or, on removing the factor ζ^{n+1-k},

$$\zeta^k = \zeta^{k-1} + \lambda \sum_{i=1}^{k} \alpha^*_{k+1,i}\zeta^{k-i} + \lambda\alpha^*_{k+1,0}\left[\zeta^{k-1} + \lambda \sum_{i=1}^{k} \alpha_{k,i}\zeta^{k-i}\right]. \qquad (8)$$

This is a polynomial equation of degree k in ζ (the characteristic equation of the method), so in general there are k distinct roots $\zeta_1, \zeta_2, \ldots, \zeta_k$. The case of fewer than k distinct roots can be handled by a limiting argument, but the conclusions are unaffected, so we ignore this case. Because (7) is linear and homogeneous, any linear combination of solutions is also a solution, hence

$$\delta_n = \beta_1\zeta_1^n + \beta_2\zeta_2^n + \cdots + \beta_k\zeta_k^n$$

is a solution of (7) for any choice of $\beta_1, \beta_2, \ldots, \beta_k$. The form of (7) clearly shows the k successive values $\delta_r, \delta_{r+1}, \ldots, \delta_{r+k-1}$ determine all succeeding values. The $\beta_1, \beta_2, \ldots, \beta_k$ are determined by the equations

$$\delta_r = \beta_1\zeta_1^r + \beta_2\zeta_2^r + \cdots + \beta_k\zeta_k^r,$$
$$\delta_{r+1} = \beta_1\zeta_1^{r+1} + \beta_2\zeta_2^{r+1} + \cdots + \beta_k\zeta_k^{r+1},$$
$$\vdots$$
$$\delta_{r+k-1} = \beta_1\zeta_1^{r+k-1} + \beta_2\zeta_2^{r+k-1} + \cdots + \beta_k\zeta_k^{r+k-1}.$$

The matrix of this set of equations is of the Vandermonde type, for which it is not difficult to find the value of the determinant [9, p. 12]. It is non-zero if the ζ_i are distinct, so there is a unique solution for the β_i, and we have uniquely determined the δ_n for all $n \geq r + k$.

In general, because any set of values of the $\delta_r, \delta_{r+1}, \ldots, \delta_{r+k-1}$ is possible, any set of values for the $\beta_1, \beta_2, \ldots, \beta_k$ is possible. For δ_n to be bounded for any set of β_i it is clearly necessary and sufficient that

$$|\zeta_i| \leq 1 \qquad i = 1, 2, \ldots, k.$$

The region of absolute stability is that region in $\lambda = hA$ with $Re(\lambda) < 0$

and all roots of (8) having a magnitude of no more than 1. For step sizes h such that λ is in the region of absolute stability the ζ_i^n do not grow; except on the boundary, they actually decrease. A lot of computation is required to locate the boundary of such a region, but it needs to be done only once for each method.

A similar analysis can be carried out for other methods. Plots are given on the following pages, for the orders $1, 2, \ldots, 12$, which compare the stability regions of the Adams-Bashforth formula of order k, the PECE method with both formulas of order k, and the PECE method with predictor of order k and corrector of order $k + 1$ (extrapolation). The stability regions for the predictor alone are generally so small that the extra evaluation of a PECE approach more than pays for itself. Extrapolation generally enhances the stability though the matter is not clear at the very highest orders. The orders 1 through 4 are the important ones for problems with stability difficulties, and extrapolation considerably increases their regions of stability. Other possibilities, such as additional corrections, have been considered but for orders $k \le 12$ the extrapolated PECE scheme is probably the most stable way to use the Adams formulas when the cost is taken into account. Reference [11] considers the matter in more detail.

STABILITY PLOTS

Stability regions for several variants of the Adams methods are displayed for orders $k = 1, 2, \ldots, 12$. Only half of each region is plotted as the regions are symmetric about the real axis. For each order k, the regions plotted are:

_____	Adams-Bashforth (predictor) formula of order k
··········	PECE method with Adams-Bashforth (predictor) formula of order k, and Adams-Moulton (corrector) formula of order k
− − − −	PECE method with Adams-Bashforth (predictor) formula of order k, and Adams-Moulton (corrector) formula of order $k + 1$. This is the method implemented in the codes provided.

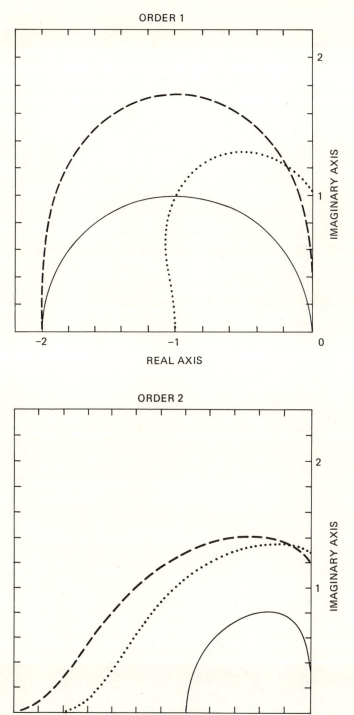

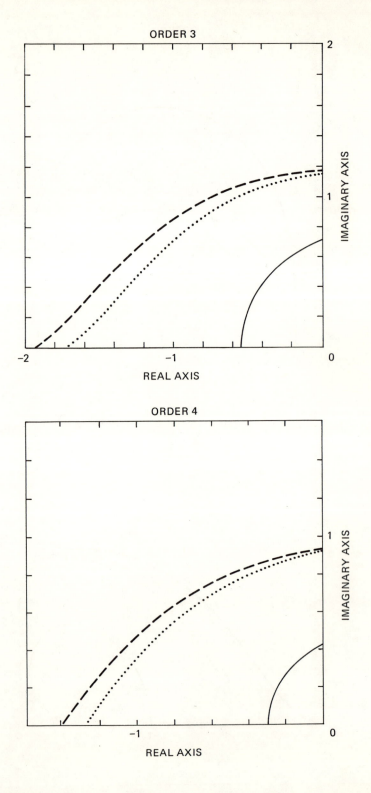

ORDER 5

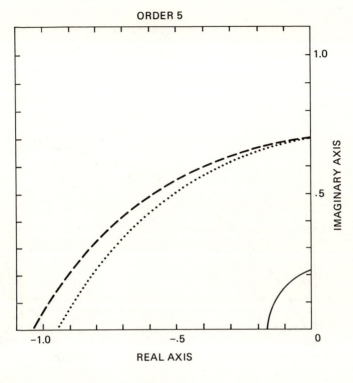

REAL AXIS

ORDER 6

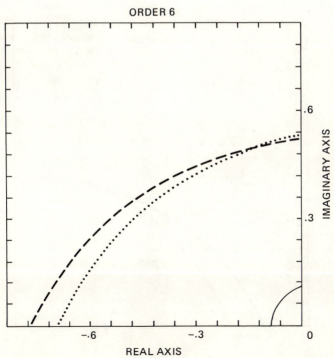

REAL AXIS

ORDER 7

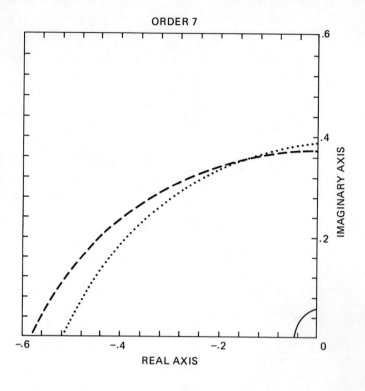

ORDER 8

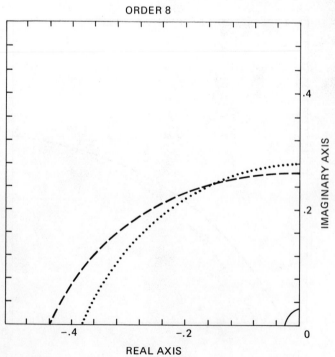

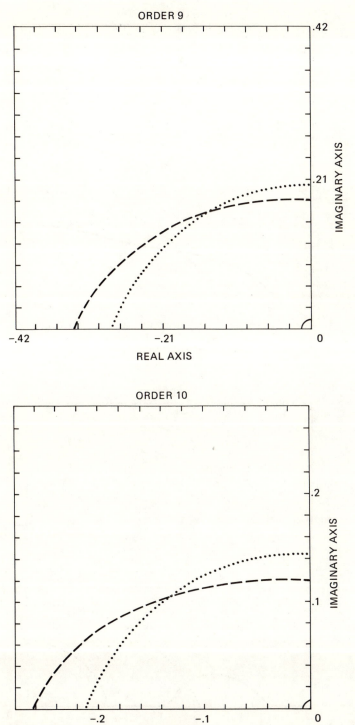

ORDER 9

IMAGINARY AXIS

REAL AXIS

ORDER 10

IMAGINARY AXIS

REAL AXIS

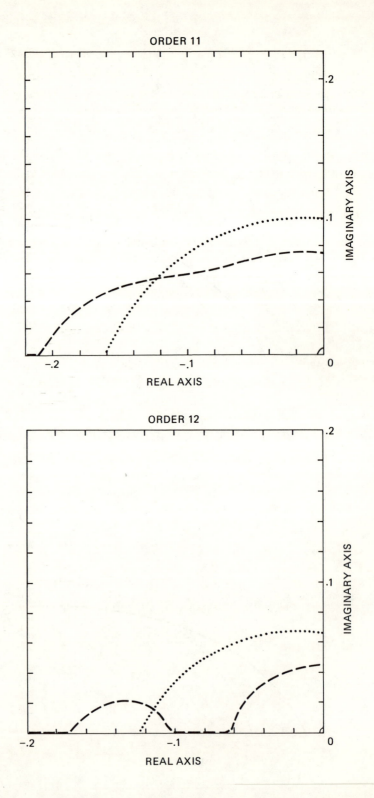

ORDER 11

REAL AXIS

IMAGINARY AXIS

ORDER 12

REAL AXIS

IMAGINARY AXIS

EXERCISE 1

What are the characteristic equations for the Adams-Bashforth formula of order k and the PECE method with both formulas of order k?

Let us again return to the role of the error estimator. For the Adams method of order k the error estimator was shown in Chapter 6 to be

$$\gamma_k^* h \nabla^k f_n^p,$$

when the step size is constant. We shall first look at the error estimate $\gamma_k^* h \nabla^k f_n$, which happens to be asymptotically equivalent but the details of its analysis are simpler. Later we shall return to the estimator used in the code. For the model problem (1) this estimate is

$$\gamma_k^* h \nabla^k \{ A[y_n + \delta_n] + g_n \} = \gamma_k^* h \nabla^k \{ A y_n + g_n \} + \gamma_k^* h A \nabla^k \delta_n.$$

Now we look at the effect of the δ_n more carefully. From the form of δ_n we see that

$$\gamma_k^* h A \nabla^k \delta_n = \gamma_k^* h A \sum_{j=1}^{k} \beta_j \nabla^k \zeta_j^n.$$

This can be clarified by writing the complex numbers $\zeta_j \neq 0$ in polar form:

$$\zeta_j = \rho_j \exp(i\theta_j).$$

Thus

$$\zeta_j^n = \rho_j^n \exp(in\theta_j)$$

and

$$\nabla \zeta_j^n = \rho_j^n \exp(in\theta_j) - \rho_j^{n-1} \exp(i(n-1)\theta_j)$$

$$= \zeta_j^n \left(1 - \frac{\exp(-i\theta_j)}{\rho_j} \right).$$

It can be seen from this that

$$\nabla^k \zeta_j^n = \zeta_j^n \left(1 - \frac{\exp(-i\theta_j)}{\rho_j} \right)^k.$$

Thus the estimated error is of the form

$$\gamma_k^* h A \nabla^k \delta_n = \sum_{j=1}^{k} \sigma_j \zeta_j^n,$$

where for each j

$$\sigma_j = \gamma_k^* h A \beta_j \left(1 - \frac{\exp(-i\theta_j)}{\rho_j} \right)^k$$

is constant.

It is only the form of the estimated error that is important to us. The estimate $\gamma_k^* h \nabla^k f_n^p$ actually used in the code can now be seen to give the same behavior. For the model problem the estimate is

$$\gamma_k^* h \nabla^k \{ A[p_n + \delta_n^p] + g_n \} = \gamma_k^* h \nabla^k \{ A p_n + g_n \} + \gamma_k^* h A \nabla^k \delta_n^p.$$

In Chapter 2 we found that

$$\nabla^k \delta_n^p = \delta_n^p + \sum_{m=1}^{k} (-1)^m \delta_{n-m} \binom{k}{m} = \delta_n^p - \delta_n + \nabla^k \delta_n.$$

Now,

$$\delta_n = \sum_{j=1}^{k} \zeta_j^n \beta_j,$$

and

$$\delta_n^p = \delta_{n-1} + \lambda \sum_{i=1}^{k} \alpha_{k,i} \delta_{n-i}$$

$$= \sum_{j=1}^{k} \frac{\zeta_j^n \beta_j}{\zeta_j} + \sum_{j=1}^{k} \zeta_j^n \left(\lambda \beta_j \sum_{i=1}^{k} \frac{\alpha_{k,i}}{\zeta_j^i} \right).$$

These expressions, and the preceding analysis of $\gamma_k^* h \nabla^k \delta_n$, show that the contribution to the error estimate of $\delta_n^p, \delta_{n-1}, \ldots$ has the form

$$\gamma_k^* h A \nabla^k \delta_n^p = \gamma_k^* h A (\delta_n^p - \delta_n) + \gamma_k^* h A \nabla^k \delta_n$$

$$= \sum_{j=1}^{k} \sigma_j^* \zeta_j^n$$

for suitable constants σ_j^*.

If $\lambda = hA$ is outside the region of absolute stability so that, for some j, $|\zeta_j| > 1$, the estimated error must grow. This growth will be detected, with the result that h will be reduced until λ is inside the region of stability. It is important to appreciate that, in any event, we estimate the error and use it to control the step size. The point of this analysis is that error due to instability is detected and properly handled automatically. Since the code works with about as large a step size as is possible, the local error is always about equal to the local error tolerance. Because of this, only a few steps are needed for the error to grow to the point that instability will be detected. When the step size is being limited by stability it will tend to oscillate about the largest stable value. This occurs because the error is actually damped out when λ is inside the region of absolute stability, and eventually the step size must be increased. If this increase carries λ outside the region of stability, there is an error growth that leads to a step size decrease, bringing λ back into the region, and so forth.

The stability plots show roughly that the stability properties of the methods decrease as the order is raised. When stiffness appears the codes move to the lower, more stable orders. There is a bias towards using low orders when they are as accurate as high orders, and this is the case for stiff problems. There is a tendency for step failures to occur as the step size fluctuates out of the stability region. The order selection algorithm tends to keep the order low since the order may be lowered, but cannot be raised, on a failed step. The order cannot be raised at all unless one has taken $k + 1$ steps with a constant step size when at order k. These facets of the order selection algorithm and the behavior of the step size make it clear that the codes will select and work with rather low orders when stiffness is present. Still other factors [16] may play a role in causing the order to be reduced in the presence of stiffness, but they are not fully understood at this time.

It is an important practical matter to have some indication that a problem has become stiff and that a different integrator should be used. The code DE has such an indicator. The idea is that low orders are rarely used except in the presence of stiffness. A sequence of 50 consecutive steps taken with orders no higher than 4 is the basic test for stiffness. Since it is possible that the low orders are due to the problem being very "easy," the test is combined with the test for too much work in DE. If the problem requires too much work (500 steps) and if there has been a long sequence of steps at low orders, DE diagnoses the trouble as stiffness. Certain misunderstandings about this test and the false returns that might result need to be discussed.

The test in DE is *not* a test for stiffness: rather, it decides whether or not stiffness is the reason for too much work. If the problem can be integrated with an acceptable expenditure of effort, it does not matter if the problem is stiff. Sometimes the problem seen by the code is stiff even though the problem posed is not. For example, the equation $y' = f(x, y) = -y^3/2$ has $f_y = -3y^2/2$. If the code is given an initial condition such that the solution $y(x)$ is "small," the problem posed is not stiff. If the code completely loses track of this solution, and begins tracking a solution that is "large," the problem seen by the code is stiff. Thus, the code may properly report stiffness even though the problem posed is not stiff. An example will be given later in this chapter.

Many stiff problems have an initial transient that is not stiff. For example, when the problem $y(0) = 1$, $y' = -100y$ is solved with a pure absolute error measure, it is usually referred to as being stiff, but there is a small interval in which the solution is decreasing rapidly and the step size is determined by accuracy rather than stability. This transient is expensive to integrate, and one might well get a return because of too much work and be properly told the problem is not stiff on this interval, depending on how far the code has gone. As the integration proceeds further the problem does become stiff, and one would expect this to be discovered. Example 14 of Chapter 11 shows this behavior.

False returns can happen. If the problem actually requires low orders and still takes too much work, a false statement of stiffness will be given. For example, suppose the solution has a severe discontinuity in a low order derivative followed by very smooth behavior. For a stringent error request, the code will drop to order one and use very small step sizes to find and pass the discontinuity. The code never raises the order until it has increased the step size to a value appropriate to the current order. In such circumstances a long sequence of steps at low orders can arise. An example of this kind is presented as Exercise 3 in Chapter 10. It is possible to detect and to account for this situation in the codes, but such problems are unusual and the test on work almost always prevents false indications. In Chapter 9 we discuss a control of the propagated roundoff error that is automatically used by the codes when an accuracy approaching the machine limit is requested. This device is justified only for non-stiff problems and its use invalidates the test for stiffness near limiting precision. In Chapter 11 there is an example of a stiff problem for which stiffness is not detected near limiting precision because of this device. There are several important assumptions in our analysis and we must expect to encounter problems for which they are invalid. Also, the bias towards low orders in the codes is not as strong as it might be and stiffness may go undetected. In particular, for stringent accuracy requirements there is a tendency to use higher orders than necessary. The following stiff problem emphasizes that the test is not perfect. For no tolerance is it reported as stiff, because no sequence of steps at low orders is long enough to pass the test. The problem is to solve, with a pure absolute error test on $[0, 20]$,

$$\mathbf{y}' = \begin{bmatrix} -1E4 & 1E3 & 0 & 0 \\ -1E3 & 1E4 & 0 & 0 \\ 0 & 0 & -1E1 & 1E2 \\ 0 & 0 & -1E2 & -1E1 \end{bmatrix} \mathbf{y}, \quad \mathbf{y}(0) = \begin{pmatrix} 1 \\ 1 \\ 1 \\ 1 \end{pmatrix}.$$

All things considered, the test works remarkably well if one keeps in mind the aspects just described. Extensive computations with standard sets of test problems have demonstrated the value of the test. We mention some results of this kind. Krogh [19] gives a set of both stiff and non-stiff problems. All were integrated with tolerances $10^{-1}, 10^{-2}, \ldots$ down to where the propagated roundoff control would be used, typically 10^{-11}. No non-stiff problem is reported as stiff. Problems 1 and 7 of this set have portions of their intervals of integration that are mildly or severely stiff. Since they do not require 500 steps, DE properly ignores this stiffness. Problem 10, when integrated with a tolerance of 10^{-1}, loses the solution curve entirely and thereby encounters stiffness, which is properly reported. For all other tolerances the code tracks the correct solution curve, leading to a non-stiff problem, and DE gives the proper return. The two stiff problems, 12 and 13, have a variety of returns. Several intervals are specified and often the code integrates over the shortest interval in less than 500 steps and, hence, does not report stiffness. There is an initial transient, so for stringent accuracy requests a return or two of too much work before reporting stiffness is common. In every case, but for the tolerance of 10^{-11} on problem 12, stiffness is detected.

9

COMPUTER ARITHMETIC

Generally speaking, one tries to ignore the difference between computer arithmetic and exact arithmetic when solving differential equations. This is usually permissible since the error introduced by the numerical method, the discretization error, is usually quite a bit larger than the error introduced by the use of limited precision arithmetic. There are, however, fundamental limitations that must be understood. In addition, some elementary precautions to ameliorate the effects of limited precision are written into the codes, and they require explanation.

Most recent numerical analysis texts discuss the fundamentals of computer arithmetic, e.g. [25]; the authoritative reference is [28]. For our purposes, it is unnecessary to develop this material. The few ideas that are needed will simply be stated and illustrated. It will be adequate to describe the floating point result of the arithmetic operations $+$, $-$, $\times$, and $/$ as the rounded exact result. The symbols $\oplus$, $\ominus$, $\otimes$, and $\oslash$ will be used for the floating point operations and $fl(x)$ will be used to denote the result of rounding x to a floating point number. Thus in five decimal digit rounded floating point arithmetic the representation of the real number $.111119 \times 10^0$ is

$$fl(.111119 \times 10^0) = .11112 \times 10^0.$$

Most computers chop rather than round—that is, they truncate all digits beyond single precision. Using chopped arithmetic,

$$fl(.111119 \times 10^0) = .11111 \times 10^0,$$

so that the interpretation of $fl(x)$ depends on the mode. We shall refer to either mode as rounding unless it is necessary to be more precise. All our numerical examples will use five decimal digits and chopping. As an example of addition, consider

$$.11111 \times 10^0 \oplus .49999 \times 10^{-3}.$$

The assumption is that $a \oplus b = fl(a + b)$, so the exact sum is

$$
\begin{array}{r}
.11111 \quad\ \times 10^0 \\
+.00049999 \times 10^0 \\
\hline
.11160999 \times 10^0
\end{array}
$$

and the chopped result is $.11160 \times 10^0$, the roundoff error being $.99900 \times 10^{-5}$. A basic technique for analyzing floating point arithmetic is to write

$$fl(a) = a(1 + \tau),$$

where τ is bounded by a unit roundoff error

$$|\tau| \le u.$$

In computers with normalized floating point arithmetic with s significant digits to the base β, numbers a are represented in the form $a = m\beta^e$, where $0 < \beta^{-1} \le m < 1$ and e is an appropriate integer. The value of u is then

$$u = \tfrac{1}{2}\beta^{1-s} \quad \text{rounded,}$$
$$= \beta^{1-s} \quad \text{chopped.}$$

Thus we can write, for example,

$$a \otimes b = fl(a \times b) = (a \times b)(1 + \tau),$$

where $|\tau| \le u$. Essentially the same conclusions are true of less idealized floating point arithmetics. The details are not important to us, and it will be quite adequate to describe u as the smallest positive value for which

$$1 \oplus u \ne 1.$$

Unless two numbers are of comparable size, the smaller will not affect their sum. For example,

$$
\begin{array}{r}
.11111 \times 10^0 \\
+ .0000049999 \times 10^0 \\
\hline
.1111149999 \times 10^0
\end{array}
$$

has the chopped result $.11111 \times 10^0$. Thus $a \oplus b = a$ if $|b| < u|a|$. This simply quantifies the statement that, if b is too small, its digits do not affect those of a in the precision used for addition. This simple observation about floating point addition poses two fundamental limitations, one on the step size and one on the local error tolerance.

A step size h is chosen so that a step can be made to $x + h$ at the required accuracy. If h is sufficiently small compared to x, then $x \oplus h = x$ and computation can proceed no further. So, the smallest step size that is meaningful has a magnitude of roughly $u|x|$. A difference between x and $x + h$ must also appear in the routine for evaluating the differential equation. Just how small the difference can be and still lead to reasonably accurate function values depends on the problem and how it is programmed. The code STEP returns control to the calling program with the variable CRASH set .TRUE. if the step size needed is less than $4u|x|$.

A similar limitation must be imposed on the tolerance. The result' of a step is of the form

$$
y_{n+1} = y_n + h \sum_{i=1}^{k} \alpha^*_{k+1,i} f_{n+1-i} + h\alpha^*_{k+1,0} f^p_{n+1}.
$$

or, as we shall abbreviate it,

$$
y_{n+1} = y_n + h \sum.
$$

The user supplies a tolerance ε and the code STEP adjusts the step size so as to make the local error of the step about 0.5ε. This presumes exact arithmetic, but there is also an error in the step due to limited precision. In particular there is a roundoff in the addition which may be as large as $u|y_{n+1}|$:

$$
y_n \oplus h \sum = (y_n + h \sum)(1 + \tau) = y_{n+1} + y_{n+1}\tau.
$$

It is pointless to attempt to control the local error to be less than the roundoff error, since it is the total error in the step which governs the accuracy.

Indeed, we need to work with tolerances of such a size that the local error is rather larger than the roundoff error in order that step changing have an effect. It is inconvenient to base a test of this kind on y_{n+1} since it is not immediately available. At limiting precision it is ordinarily the case that $|y_n| \gg |h\sum|$, so $|y_{n+1}| \doteq |y_n|$. The code STEP tests if

$$0.5\varepsilon < 2u\,|y_n|$$

when it is called. If the inequality holds, the code increases ε until $0.5\varepsilon \doteq 2u\,|y_n|$ and returns control to the calling program with CRASH = .TRUE.. Detecting requests for too much accuracy in this way costs no function evaluations and users are informed of the smallest ε that they can reasonably expect to use.

This limitation on the tolerance is very important when one accidentally asks for too much accuracy because he does not know enough about the solution or because he wants all the accuracy possible. One might think that the limitation on the step size would suffice, but it does not. It is very common to find that, while codes of this kind appear to work normally when the tolerance is reduced below this limit, they actually return worse answers at greatly increased expense. As we shall see in Chapter 11 this simple test works quite well in getting nearly the greatest possible accuracy while also controlling the cost.

These two tests prevent the codes from attempting to obtain accuracies that are unreasonable because of the machine precision. Let us now examine some devices for increasing the accuracy which is possible in a given precision. It is repeated addition (subtraction) that leads to dangerous error growth in floating point arithmetic. A rule of thumb [28, pp. 18–19] for controlling the error is to add numbers in roughly the order of increasing magnitude. We can arrive at this rule by writing out an expression for the error, using the assumptions made earlier, and by observing that the rule tends to minimize a natural bound on the error. Sums involving differences are repeatedly formed, as in the evaluation of p_{n+1} and p'_{n+1}, and in interpolation. A simple example is

$$p'_{n+1} = \sum_{i=1}^{k} \phi_i^*(n).$$

There are several reasons for thinking the differences in these sums tend to decrease as their order increases. We shall examine these reasons in a moment but, assuming for now that they do decrease, the rule of thumb suggests summing them in reversed order, that is, by

$$\sum \leftarrow \phi_k^*(n) \qquad \text{initialize,}$$

$$\sum \leftarrow \sum + \phi_i^*(n) \qquad i = k - 1, k - 2, \ldots, 1,$$

$$p'_{n+1} \leftarrow \sum.$$

This action has an effect which is usually beneficial (though small) and if the differences do not decrease, no harm is done. Since it is as easy to sum in reversed order as not, the device is employed throughout the codes.

The differences used by the codes tend to decrease as their order increases. The representation of differences as scaled derivatives (see Chapter 2) shows that the differences of an interpolating polynomial are, in a rough way, analogous to coefficients in a convergent Taylor series expansion of the solution, and so they also tend to decrease. The error of the numerical method of order k is estimated by a multiple of a difference, $h\gamma_k^* \nabla^k f_n^p$ in the case of constant step size. The multiple does not vary much as a function of the order, so that it is the use of an order k yielding small differences which allows one to attain high accuracy. The algorithm for controlling the order roughly adds or drops differences so that they tend to decrease up to the maximum order used, and to increase for higher orders. In the important case of roundoff being troublesome, the step size is usually small and the function values nearly constant. The differences are then formed by subtraction of nearly equal quantities, and so they decrease rapidly. Finally, examination of difference tables for representative test problems shows a strong tendency for the differences to decrease up to some order and then to increase. The fact that they tend to decrease to the order chosen by the code justifies the rule of thumb employed.

There is an effect causing differences to increase which we now want to examine. In order to be specific, suppose that a constant step size h is being used. In Chapter 2 we found that

$$\nabla^k f_n = \sum_{m=0}^{k} (-1)^m f_{n-m} \binom{k}{m}.$$

Because errors are made in the routine for evaluating f_n we actually have

$$f_n^* = f_n(1 + \delta_n) = f_n + f_n \delta_n,$$

for some relative error δ_n. Then

$$\nabla^k f_n^* = \nabla^k f_n + \sum_{m=0}^{k} (-1)^m f_{n-m} \delta_{n-m} \binom{k}{m}.$$

The errors δ_n are usually small, but they can be amplified by differencing. The growth can be bounded by

$$\left| \sum_{m=0}^{k} (-1)^m f_{n-m} \delta_{n-m} \binom{k}{m} \right| \le \sum_{m=0}^{k} \binom{k}{m} \cdot \max_{0 \le m \le k} |f_{n-m} \delta_{n-m}|$$

$$= 2^k \max_{0 \le m \le k} |f_{n-m} \delta_{n-m}|.$$

(The simple expression comes from a well known identity,

$$2^k = (1+1)^k = \sum_{m=0}^{k} \binom{k}{m}.)$$

If the f_{n-m} are constant (as they nearly are at limiting precision) and the δ_{n-m} are constant but of alternating sign, the bound is achieved. If the errors δ_n are thought of as random variables with their sign just as likely to be negative as positive, then an average value for the error in the difference is zero. This is because for each error in the difference of the form

$$\sum_{m=0}^{k} (-1)^m f_{n-m} \delta_{n-m} \binom{k}{m},$$

there will be a corresponding one of the form

$$\sum_{m=0}^{k} (-1)^m f_{n-m} (-\delta_{n-m}) \binom{k}{m},$$

which leads to an average value of zero. In a vague, "average" sense we expect that the amplification of the errors in the f_n will be moderate but it can be large in the higher differences. The order selection procedure is affected by the fact that it is $\nabla^k f_n^*$ which is used. We anticipate that the true differences $\nabla^k f_n$ will decrease as a function of the order k. However, because of amplification of the errors in evaluating the f_n, we should expect that the differences $\nabla^k f_n^*$ will actually decrease up to a certain order and then increase as the roundoff effects dominate the discretization error associated with the true differences. This expectation of the qualitative behavior is amply borne out by experience. The order selection process tends to stop using differences when they start to increase, thus controlling the order so as to use only those differences with some meaningful figures.

An important source of error at limiting precision is in the additions in the main computational steps

$$p_{n+1} = y_n + h_{n+1} \sum_{i=1}^{k} g_{i,1} \phi_i^*(n) = y_n + h_{n+1} \sum_{i=1}^{k} \alpha_{k,i} f_{n+1-i},$$

$$y_{n+1} = p_{n+1} + h_{n+1} g_{k+1,1} [f_{n+1}^p - p'_{n+1}].$$

To expose the important additions and to simplify the notation, we shall write these formulas as

$$p_{n+1} = y_n + h \sum f_n$$

and

$$y_{n+1} = p_{n+1} + h S_p f_{n+1}.$$

All that we need $\sum f_n$ for is to indicate a linear combination of values f_{n+1-i}, and $S_p f_{n+1}$ a similar combination with f_{n+1}^p, too. We cannot reduce the addition errors below a unit roundoff when they arise, but we shall attempt to control their effect on future steps.

There is an alternative to the approach that we take, which uses some double precision while working in single [2, p. 327]. We do not develop this technique because we expect the user to be working in full double precision when this is reasonable. Let us discuss the matter more fully.

Typical machines that have been used in conjunction with this text are an IBM 360/67, a PDP-10, and a CDC 6600. In single precision the IBM 360 systems are base 16 machines with chopped arithmetic, carrying about 7 decimal digits. The word length and roundoff properties are so poor that single precision is all but useless for the serious solution of differential equations (as distinguished from solutions done for classroom illustration). On the other hand, the double precision of about 16 decimal digits is built into the hardware and its use costs only about 25% more than single precision. For these reasons one should ordinarily use full double precision to solve differential equations on this line of machines. (Some marvel at how cheap double precision is, others at why such a poor single precision was provided.) The PDP-10 is a base 2 machine with rounding and about 8 decimal digits. Because the double precision is supplied by software it is relatively expensive. The rounded binary arithmetic is very helpful near limiting precision. With the roundoff controls that are built into our codes, single precision will usually suffice, but high precision computation will require the use of double precision. The CDC 6600 is a base 2 machine with chopping and about 14 decimal digits. It would be quite unusual to want to use double precision on this machine. (This is fortunate, since it is provided by software.) In summary, we anticipate users of IBM 360 system machines to be working in full double precision, and PDP-10 and CDC 6600 users to be working in single precision. In any case it is quite expensive to increase the precision.

Let us now return to the question of propagated roundoff error and discuss how to reduce it without going to higher precision. The values we wish to compute are

$$p_{n+1} = y_n + h \sum f_n$$

and

$$y_{n+1} = p_{n+1} + hS_p f_{n+1}.$$

Let y_n^*, p_n^*, etc. be the values actually used in the computation. They have small errors associated with them, due to the cumulative effect of floating point arithmetic. By rearranging the computation we shall attempt to keep the errors, $\delta_n = y_n^* - y_n$ and $\delta_n^p = p_n^* - p_n$, down to the size of a unit roundoff. The basic approach involves estimating just the dominant terms contributing to the propagated error. The essential ingredient is a set of realistic assumptions about the relative sizes of the computed quantities.

The error propagated into p_{n+1} comes from three sources: those errors already in y_n, those errors resulting from the addition of $h \sum f_n$ to y_n, and those resulting from the formation of $h \sum f_n$. By the device to be employed, the error in y_n will be of the order of a unit roundoff. When roundoff is important the step size is usually small enough so that the solution changes little from step to step, $y_n \doteq p_{n+1} \doteq y_{n+1}$. This implies that $|y_n| \gg |h \sum f_n|$. Therefore, the roundoff resulting from the addition of the two terms to form p_{n+1} is of the order of a unit roundoff in $p_{n+1} \doteq y_n$. The error in the formation of $h \sum f_n$ has several origins. The terms f_i^* actually used have errors coming from two sources. One is due to the use of y_i^* instead of y_i. This error satisfies

$$f(x_i, y_i^*) = f(x_i, y_i + \delta_i) = f_i + f_y \delta_i + \cdots = f_i + O(uf_y).$$

(Here, as elsewhere in this discussion, we use the symbol $O(uf_y)$ informally to mean quantities of the general size or smaller than uf_y.) We presume that the routine for evaluating f yields results accurate to a few units of roundoff, so that

$$f^*(x_i, y_i^*) = f(x_i, y_i^*) + O(uf_i) = f_i + O(uf_i) + O(uf_y).$$

The arithmetic errors in forming $h \sum f_n$ should normally amount to no more than a few units of roundoff in the result, so that

$$(h \sum f_n)^* = h \sum f_n + O(uh \sum f_n) + O(uhf_y).$$

We are interested in the high precision solution of non-stiff problems, in which case h will normally be small enough so that we are well within the region of absolute stability, i.e., $1 \gg |hf_y|$. If the solution is such that $|y_n| \gg |hf_y|$, then we can see that the errors in the formation of $h \sum f_n$ are all small compared to the other two sources of error in p_{n+1}.

Using an estimate δ_n^* of δ_n we shall attempt to correct the predicted value, but this must be done carefully. By its construction δ_n^* is of the order of a unit roundoff in y_n so simply subtracting it from p_{n+1}^* causes a roundoff error as big as the intended correction. However, the formation of

$$\tau = (h \sum f_n)^* \ominus \delta_n^*$$

does the correction accurately, because the two terms are of about the same size and the addition error of a unit roundoff of τ will be negligible in comparison to the addition error ξ in

$$p_{n+1}^* = y_n^* \oplus \tau.$$

In summary, the error δ_{n+1}^p in

$$p_{n+1}^* = p_{n+1} + \delta_{n+1}^p = y_n^* + \tau + \xi$$
$$= y_n + \delta_n + (h \sum f_n)^* \ominus \delta_n^* + \xi$$
$$\doteq y_n + \delta_n + h \sum f_n - \delta_n + \xi,$$

is very nearly equal to the addition error ξ. It can be determined accurately from

$$\delta_{n+1}^p \doteq \xi = (p_{n+1}^* \ominus y_n^*) \ominus \tau.$$

This is possible because $p_{n+1}^* \doteq y_n^*$ so that their leading digits cancel in the subtraction and there is no roundoff error at all. Similarly there is no roundoff error in the second subtraction.

If we proceed in the same way to correct the error δ_{n+1}^p in the computation of y_{n+1}^* and to estimate the propagated error δ_{n+1}, we arrive at

$$\rho = (hS_p f_{n+1})^* \ominus \xi,$$

$$y_{n+1}^* = p_{n+1}^* \oplus \rho,$$

and

$$\delta_{n+1} \doteq (y_{n+1}^* \ominus p_{n+1}^*) \ominus \rho.$$

This completes a typical step. It is important that we attempt to account for both δ_n and δ_{n+1}^p at each step so that they are kept to the order of a unit roundoff since this assumption is basic to the justification offered.

The assumptions made are certainly not valid when the solution passes through a zero, and are unlikely to be valid unless one is approaching limiting precision. The computation as rearranged would have no effect at all if the machine arithmetic were exact, and in any event it alters the individual values by only a few units of roundoff. Unless one is at limiting precision and taking many steps, the effect should be invisible. If the assumptions are valid, the effect will still depend on how error propagates for the particular problem. Since there also is a complex interaction with the step and order selection, it is difficult to predict exactly what will happen. Although it may seem surprising, the result is likely to be fewer function evaluations. This is because the differences are more accurate. As a result the error estimators function more effectively for reasons examined earlier, and higher order formulas and larger step sizes are seen to be permissible. The gains in efficiency and accuracy due to this device are quite modest but our code is intended for the high precision solution of non-stiff problems so we have incorporated it exactly as described. Numerical experiments on representative test problems found the effect of the device to be invisible at tolerances greater than about 100 times limiting precision. To reduce the overhead, the device is not used if the initial tolerance is greater than 100 times limiting precision on the first call to the code.

To fully analyze roundoff effects in a differential equation solver is quite beyond the state of the art. In Chapter 11 we shall examine some experimental evidence to complement the theory of this chapter. On the whole the codes provided behave very satisfactorily with respect to limiting precision. This aspect of differential equation solvers has been rather neglected and there is little written about work in the area, though there is a very interesting body of theory based on a statistical model of roundoff [2, 12].

10

THE CODES

In this chapter we provide subroutines for solving ordinary differential equations that use the algorithms developed in the preceding chapters. The main codes are the integrator STEP, the interpolation routine INTRP, and the driver DE. The codes STEP and INTRP advance the numerical solution of an equation one step, and interpolate the solution and its derivative, respectively. The theory supporting these codes and the implementation of the algorithms have been explained in earlier chapters, so we will just summarize the main ideas here. The subroutine DE is a driver that simplifies using the other codes for routine problems; it will be explained at length. We have attempted to make the codes as easy as possible to use, as well as efficient and robust.

We first discuss how to implement and test the codes on various computing systems. Each code has a prologue of comments explaining how to use it; these comments are expanded here and supplemented with examples. In the third section we shall explain the codes. We have attempted to program them so as to be readable and efficient. The structuring, disciplined use of FORTRAN, and devices for efficiency will be explained. The codes are extensively commented and, where appropriate, we supply flow charts. Next we discuss extensions of the codes that increase their usefulness but that also make them harder to use or less portable. A root-solving code is given in the last section along with instructions for its use.

Advances in the theory, and the experience of users, may prompt alterations to these codes. Readers are invited to write the authors for the changes that they endorse.

IMPLEMENTING THE CODES

Four basic codes are provided: the core integrator STEP, an interpolation routine INTRP, a driver DE, and a subroutine MACHIN for calculating the unit roundoff of the machine being used. We list single precision versions because they are more portable and more appropriate for general use on large computers.

The only machine dependent constants are TWOU and FOURU, which represent, respectively, two times and four times the machine's unit roundoff error U. This latter quantity is defined as being the smallest positive number for which $1 \oplus U > 1$. Because the constants depend on the machine and the precision being used, they must be calculated and inserted in DATA statements in both STEP and DE before using these codes. If the base β and the number of significant digits s (the word length) of the machine are known, U can be calculated from the relations given in Chapter 9,

$$U = \beta^{1-s} \qquad \text{chopped arithmetic,}$$
$$= \tfrac{1}{2}\beta^{1-s} \qquad \text{rounded arithmetic.}$$

If this information is not available, the subroutine MACHIN will calculate U from its definition. The constants TWOU and FOURU need have only one or two correct digits. For the following computers U is approximately:

IBM 360/67	single precision	9.5E-7,
UNIVAC 1108	single precision	1.5E-8,
PDP-10	single precision	7.5E-9,
CDC 6600	single precision	7.1E-15,
IBM 360/67	double precision	2.2D-16.

We have used ANS standard FORTRAN in all but two situations. Several variables and arrays are stored in STEP and DE between calls and are not passed through the call list. This procedure is common to integrators like ours because it keeps the call list down to a convenient length. It can cause trouble in an overlay situation or when one wants to switch back and forth between problems without restarting. These situations are not common to the classroom and are somewhat unusual in industrial laboratories. The resolution of these difficulties is easy enough for most compilers and we discuss how to do this in another section, along with other valuable extensions of the codes. Unfortunately, the WATFOR compiler, which is widely used in academic computer centers, does not allow one of the simple extensions we give, even though it is standard FORTRAN. For this reason the basic codes use the non-standard, but common, FORTRAN device of internal storage. We also employ a non-standard DATA statement because the 1965 FORTRAN standard is very restrictive. The form used is acceptable on all machines known to us and it is expected to become standard.

Double precision versions of these codes require several modifications. All real variables must be declared double precision; this can be done easily by declaring

$$\text{IMPLICIT REAL*8 (A-H,O-Z)}$$

on IBM machines and

$$\text{IMPLICIT DOUBLE PRECISION (A-H,O-Z)}$$

on many others. On machines with FORTRANs not implementing this feature, like the CDC 6600, all variables must be listed in a type declaration. All constants must be converted to double precision. The constants stored in DATA statements require only a few digits of accuracy, so they are already sufficiently accurate for double precision use. FORTRAN-supplied functions like ABS must be changed to their double precision counterparts. On machines with FORTRANs not implementing the IMPLICIT type declaration, it may be necessary to explicitly declare these functions as double precision. For the convenience of the user who must declare all the variables double precision by listing them, we supply a list of the variables and FORTRAN functions in each of the basic routines:

DE—ABSDEL, ABSEPS, ABSERR, DEL, DELSGN, EPS,
    FOURU, H, HOLD, P, PHI, PSI, RELEPS, RELERR,
    T, TEND, TOLD, TOUT, WT, X, Y, YP, YPOUT, YY
    Functions—ABS, AMAX1, AMIN1, SIGN

STEP—ABSH, ALPHA, BETA, EPS, ERK, ERKM1, ERKM2,
    ERKP1, ERR, FOURU, G, GSTR, H, HNEW, HOLD,
    P, PHI, PSI, P5EPS, R, REALI, REALNS, RHO,
    ROUND, SIG, SUM, TAU, TEMP1, TEMP2, TEMP3,
    TEMP4, TEMP5, TEMP6, TWO, TWOU, V, W, WT,
    X, XOLD, Y, YP
    Functions—ABS, AMAX1, AMIN1, SIGN, SQRT

INTRP—ETA, G, GAMMA, HI, PHI, PSI, PSIJM1, RHO,
    TEMP1, TEMP2, TEMP3, TERM, W, X, XOUT, Y,
    YOUT, YPOUT
    Functions—none

MACHIN—HALFU, TEMP1, U
    Functions—none

Once the codes have been implemented, they should be tested to verify that they are working properly. Subroutines INTRP and MACHIN are so short and straightforward that they can be verified by inspection. STEP and DE are sufficiently long and complicated that inspection must be supplemented with some computation. The examples of this chapter and the next can be used for this testing. Obviously we cannot exactly duplicate the numbers on a different machine, but the extensive performance figures of the next chapter and especially those problems exercising all of DE and STEP will give some confidence that the codes have been properly implemented. There are few constants to be checked. One must be very careful to use suitable values of TWOU and FOURU in both DE and STEP. STEP has two arrays of constants. The powers of 2 stored in the vector TWO are trivially checked. The constants γ_i^* can be generated by a simple recursion [3] for testing purposes, $\gamma_0^* = 1$,

$$\gamma_m^* + \frac{1}{2}\gamma_{m-1}^* + \frac{1}{3}\gamma_{m-2}^* + \cdots + \frac{1}{m+1}\gamma_0^* = 0, \qquad m = 1, 2, \ldots.$$

They are stored as $\mathrm{GSTR}(I) = |\gamma_{I-1}^*|$, for $I = 1, 2, \ldots$.

The storage requirement for STEP is considerable, roughly $21 \cdot \mathrm{NEQN}$ words where NEQN is the number of equations. A moderately large system of equations may cause the memory requirement to exceed the capacity of a small machine. The same difficulty will occur on any machine

for a sufficiently large system. The first action to be taken in easing this difficulty is to eliminate the propagated roundoff control; this saves $2 \cdot$ NEQN storage locations. Only the following modifications are needed:
Declarations:
change the DIMENSION statement to

```
    DIMENSION  Y(NEQN),WT(NEQN),PHI(NEQN,14),P(NEQN),
  1    YP(NEQN),PSI(12)
```

change the LOGICAL statement to

```
            LOGICAL  START,CRASH,PHASE1
```

Block 0:

```
        delete
        NORND = .TRUE.
        IF(P5EPS .GT. 100.0*ROUND) GO TO 99
        NORND = .FALSE.
        DO 25 L = 1,NEQN
 25        PHI(L,15) = 0.0
```

Block 2:

```
        delete
        IF(NORND) GO TO 240
        DO 235 L = 1,NEQN
         TAU = H*P(L) − PHI(L, 15)
         P(L) = Y(L) + TAU
235        PHI(L,16) = (P(L) − Y(L)) − TAU
        GO TO 250
```

Block 4:

```
        delete
        IF(NORND) GO TO 410
        DO 405 L = 1,NEQN
         RHO = TEMP1*(YP(L) − PHI(L,1)) − PHI(L,16)
         Y(L) = P(L) + RHO
405        PHI(L,15) = (Y(L) − P(L)) − RHO
        GO TO 420
```

If this reduction is not sufficient, the maximum order of the approximations can be lowered. For example, restricting the maximum order to $k = 6$ saves another $6 \cdot$ NEQN locations. Besides these changes, the following modifications are needed:

Declarations:
change the DIMENSION statements to

    DIMENSION Y(NEQN),WT(NEQN),PHI(NEQN,8),P(NEQN),
    1   YP(NEQN),PSI(6)
    DIMENSION ALPHA(6),BETA(6),SIG(7),W(6),V(6),G(7),
    1   GSTR(7),TWO(7)

and reduce the DATA statements for GSTR and TWO to their first seven entries.

Block 4:

    replace
       IF(KNEW.EQ.KM1 .OR. K.EQ.12) PHASE1 = .FALSE.
    with
       IF(KNEW.EQ.KM1 .OR. K.EQ.6) PHASE1 = .FALSE.
    replace
       IF(ERKP1.GE.ERK .OR. K.EQ.12) GO TO 460
    with
       IF(ERKP1.GE.ERK .OR. K.EQ.6) GO TO 460

Several arrays in the driver DE are of absolute dimension 20, thus limiting to 20 the number of equations it can integrate. To change this, only the dimensions of 20 in the statement

    DIMENSION YY(20),WT(20),PHI(20,16),P(20),YP(20),YPOUT(20)

and the number 20 in the first executable statement need be changed. An alternative implementation allowing variable dimensions is discussed later in this chapter.

USING THE CODES

We discuss using the three codes STEP, INTRP, and DE in the order that a typical user will need them. The subroutine DE is simply a driver calling STEP and INTRP; most users will find it easy to use and quite satisfactory for routine problems. Subroutine STEP is the basic Adams integrator. Used alone it allows great flexibility in solving a problem but as a result demands more of the user than does DE. Most people will use STEP only when DE is not sufficiently flexible to handle their requirements.

Subroutine INTRP is an interpolation code for obtaining the solution at specified output points. It cannot stand alone and must be used in conjunction with STEP.

Both STEP and DE use Adams methods, which are most advantageous for problems in which function evaluations are expensive, or when moderate to high accuracy is requested, or when many output points are required. The overhead in these codes is rather high, partly because of the methods and partly because of the features they provide. For example, DE detects and deals with discontinuities; it detects and deals with moderate stiffness; it detects and tells the user of severe stiffness; it detects requests for very high accuracy and switches on propagated roundoff controls; it detects requests for more accuracy than is possible on the machine being used and tells the user what *is* possible; it is, for practical purposes, independent of the number and location of output points; it monitors the work being expended; it allows the user to change direction without restarting; and it is extremely efficient in terms of function evaluations. If each function evaluation is very cheap, so that the total number is not very important, or, if the requested accuracy is low, so that roundoff precautions and high orders are superfluous and step sizes are already large, then Adams methods are not the most efficient. However, under these conditions the computation is usually cheap for any method.

DE: Few differential equations arise as the single equation

$$y'(t) = f(t, y(t)),$$

$$y(a) = y_0$$

as we assumed earlier in the text. Rather they normally appear as a system of first order equations

$$y_1'(t) = f_1(t, y_1(t), y_2(t), \dots, y_n(t)),$$
$$y_2'(t) = f_2(t, y_1(t), y_2(t), \dots, y_n(t)),$$
$$\vdots \qquad \vdots \qquad \qquad \qquad \qquad (1)$$
$$y_n'(t) = f_n(t, y_1(t), y_2(t), \dots, y_n(t))$$

with initial conditions

$$y_1(a), y_2(a), \dots, y_n(a) \qquad \text{given.} \qquad (2)$$

These systems occur naturally or arise from higher order equations as explained in Chapter 1. Usually equations (1) and (2) are written

$$\mathbf{y}'(t) = \mathbf{f}(t, \mathbf{y}(t)),$$

$$\mathbf{y}(a) = \mathbf{y}_0$$

to correspond to the notation for the single equation.

The ideal situation for anyone solving the problem posed in (1) and (2) numerically is to obtain the solutions $y_1(b), y_2(b), \ldots, y_n(b)$ at any point b he chooses, as accurately as he chooses, with the least amount of effort on his part. The minimum amount of information that anyone can expect to supply any integrator is a complete specification of the problem. This involves defining the equations (usually in the form of a subroutine for evaluating them) and the initial conditions $y_1(a), y_2(a), \ldots, y_n(a)$; indicating the interval of integration $[a, b]$; and declaring what accuracy is expected and how the error is to be measured. The integrator should report the solutions at b and the fact that it succeeded; or, if it does not complete the integration, the solutions at the place that it failed and why it failed. Subroutine DE very nearly satisfies these ideal, minimal requirements.

DE is intended to be used for problems in which only the solutions at the endpoint b are of interest or, more commonly, when a table of the solutions at a sequence of output points is desired. The user receives no information regarding the integration within $[a, b]$, so DE is inappropriate for special problems in which the solution, order, step size, etc., are to be monitored during the integration. In such cases users must write their own drivers for STEP.

All essential information is passed to and from DE with only eight parameters:

SUBROUTINE DE(F,NEQN,Y,T,TOUT,RELERR,ABSERR,IFLAG)

NEQN represents the number of equations n in (1); it cannot exceed 20 unless a DIMENSION statement is modified (see the first section) or a different version of the routine is used (as discussed below). F is the name of a subroutine defining the differential equations. It must be declared in an EXTERNAL statement in the calling program and be of the form

SUBROUTINE F(T,Y,YP)

where Y and YP are vectors dimensioned (at least) NEQN in the subroutine. F evaluates

$$\text{YP(L)} = f_L(\text{T,Y(1),Y(2)}, \ldots, \text{Y(NEQN)})$$

for $L = 1, 2, \ldots, \text{NEQN}$, that is, it evaluates the right hand side of (1). The vector Y is dimensioned NEQN in the calling program. On input it is the vector of initial conditions,

$$\text{input}: \ Y(L) = y_L(a), \qquad L = 1, 2, \ldots, \text{NEQN}.$$

After a successful computation it is

$$\text{output}: \ Y(L) = y_L(b), \qquad L = 1, 2, \ldots, \text{NEQN}.$$

The variables T and TOUT define the interval of integration, i.e., $T = a$ and $\text{TOUT} = b$. It does not matter if $b < a$. The parameters Y, T, RELERR, ABSERR, and IFLAG must be variables in the calling program since DE changes their values.

The user supplies subroutine DE with the non-negative relative and absolute error tolerances RELERR and ABSERR. The code attempts at each internal step to control each component of the local error vector so that

$$|\text{local error}_L| \leq \text{RELERR}*|Y(L)| + \text{ABSERR}.$$

This is a mixed relative-absolute error criterion that includes, as special cases, pure absolute error ($\text{RELERR} = 0$) and pure relative error ($\text{ABSERR} = 0$). For reasons discussed in Chapter 6, the code does not attempt to control the global error directly, so it is not necessarily true that

$$|y_L(b) - Y(L)| \leq \text{RELERR}*|Y(L)| + \text{ABSERR}$$

at the end of the integration. For most practical problems this inequality is approximately true and one chooses RELERR and ABSERR accordingly. As discussed in Chapter 7, a reliable way to estimate the global error with this code is to repeat the integration with the tolerances reduced, for example, by a factor of 0.1. The smaller tolerances lead to more accurate results so that the accuracy of the first integration can be estimated by comparison. An alternative way of assessing the quality of the computed solution is discussed in Chapter 7, and later in this chapter we describe how to modify DE so as to implement it.

It is up to the user to choose an appropriate error criterion since the code cannot judge how suitable the specified measure is for a given problem. An unreasonable choice results in wasted computation, and it can result in error returns from the code. Pure relative error is not defined when the solution vanishes. Since the code approximates the solution throughout the interval $[a, b]$, this measure is not defined if the solution vanishes *anywhere* in the interval and an attempt to use it may result in a divide check or an overflow. In practice the only situation at all likely to cause this difficulty is if one of the initial values is exactly zero. It is easy to alter DE to detect misuse of this sort but it involves the irritation of requiring an extra value for a flag and extra overhead for all users, in order to spot what is nearly always a perfectly obvious oversight. In the section on extensions of the codes there are instructions for making the code more robust in this respect if one wishes to do so. The only remedy is to use a non-zero absolute error tolerance which is "small," depending on the scale of the problem, over a short interval containing the zero of the solution. Pure absolute error may be impossible if the solution becomes very large, depending on the accuracy requested and the machine being used, and the code may return with a message to this effect.

We suggest the following as rules of thumb for choosing a criterion. If the solution changes a great deal in magnitude during the integration, and you wish to see this change, use relative error. If the solution does not vary much, or if you are not interested in it when it is small, use absolute error. A mixed criterion is probably the best and safest choice. For solutions large in magnitude it is essentially relative error and for solutions small in magnitude it is essentially absolute error. Thus it is always a reasonable choice and avoids the troubles of the pure criteria.

This completes the specification of the problem. We now turn to the way in which DE reports success or failure through the flag IFLAG. Normal return has

> IFLAG = 2—integration successful, T is set to TOUT and Y to the solution at TOUT.

All parameters in the call list are set for continuing the integration if the user wishes to. All he has to do is define a new value TOUT and call DE again.

There is a control on the accuracy requested:

> IFLAG = 3—error tolerances RELERR and ABSERR are too small for the machine being used. T is set to the point closest to TOUT reached during the integration and Y to the solution at that point. RELERR and ABSERR are set to larger, acceptable values. To continue with the larger tolerances, just call DE again.

The word length of the machine and the error criterion impose limitations on the accuracy that can be obtained. Requests for too much accuracy are detected without additional function evaluations (calls to the subroutine F) and suitable tolerances are communicated to the user. Because of the way control is returned, nothing is lost. The user gets the last solution computed by the code which meets the requested error tolerances and, if he wishes to continue with the larger tolerances, he simply calls DE again. To integrate with the maximum accuracy possible, simply specify tolerances that are known to be too small and let the code increase them to an acceptable level. The code will increase the tolerances if it is necessary, but will not decrease them. The tolerances may be altered by the user at each call without re-initializing.

There is a control on the work allowed:

IFLAG = 4—more than MAXNUM steps are required to reach TOUT. T is set to the point closest to TOUT reached during the integration, and Y to the answer at that point. To continue, just call DE again.

Work is measured by the number of steps taken internally; this involves about two calls to the subroutine F for each step. An arbitrary limit MAXNUM = 500 is imposed. This is a lot of computing for an efficient code but is still a reasonable cost for most problems. Just as in requesting too much accuracy, nothing is lost. The integration up to T has been successful and if the user is willing to spend more to continue, everything is set for calling DE again. An appropriate value for MAXNUM depends on the number and complexity of the equations being solved, the length of the interval of integration, the error tolerances, and the user's intentions in solving the problem. MAXNUM could be included in the call list, to allow greater flexibility, but this is inconvenient for the casual user, who would probably supply an equally arbitrary value anyway. For problems where 500 steps cost too much, MAXNUM can be decreased by altering the DATA statement in which it is defined; where 500 steps are insufficient, the user can increase MAXNUM or simply call DE again.

There is also an indicator for stiff equations:

IFLAG = 5—more than MAXNUM steps needed to reach TOUT and the equations appear to be stiff. T is set to the point closest to TOUT reached during the integration, and Y to the answer at that point. A code for stiff equations should be used but one can (usually) get accurate results with DE if he is prepared to stand the cost. To continue, the user has only to call DE again.

The nature of the test for stiffness is discussed in Chapter 8 and again in connection with an example in the next chapter. The test is adequate if it is properly used but we emphasize that it is not perfect. Example 14 in Chapter 11 and the relevant discussion should be read to understand this indicator, and the reader should also look at Exercises 2 and 3.

All parameters are checked for proper sign and range:

IFLAG = 6—integration is not begun because the input parameters are invalid. The user must correct them and call DE again.

The following input is *valid*:

$1 \leq$ NEQN ≤ 20;
T $\neq$ TOUT;
RELERR ≥ 0;
ABSERR ≥ 0;
RELERR and ABSERR not both zero;
$1 \leq |$IFLAG$| \leq 5$;
when continuing an integration, the input value of T is the output value of TOUT from the preceding call.

The last condition is automatically set by DE and this test verifies that the user did not accidentally alter T when setting a new TOUT.

So far DE is an "ideal" integrator, requiring only that the user specify his problem and returning information as to the success or failure of the integration. There are two other pieces of information that the user must supply through IFLAG. The code must be told of the first call so that it can initialize itself. This is done by an input value IFLAG = 1. The code is organized so that the output value of IFLAG is appropriate for continuing the integration. To go on to another output point the user need only define TOUT appropriately. If he wishes, he may alter the tolerances RELERR and ABSERR but he should not alter the other parameters. In particular it is important *not* to reset IFLAG to 1 as this will cause the code to restart itself. Besides being inefficient, doing this leads to less accurate results. If the user starts on another problem, he must of course restart. Reversing the direction of integration constitutes a new problem, but we found that this fact was so frequently forgotten by users that we included a check in DE for reversed direction. If the user forgets to re-initialize the code in this situation, it is done automatically, so the only time the user must re-initialize is when changing equations. An example of this, which is not obvious, occurs when one integrates

$$\mathbf{y}' = \mathbf{f}(x, \mathbf{y}), \mathbf{y}(a) = \mathbf{y}_0$$

on $[a, b]$ and uses $\mathbf{y}(b)$ as initial conditions for

$$\mathbf{z}' = \mathbf{g}(x, \mathbf{z}), \mathbf{z}(b) = \mathbf{y}(b)$$

to be integrated on $[b, c]$. The second problem is a new one to the code and therefore the code must be restarted.

For most problems this completes a description of how to use DE. There is a potential source of difficulty which is dealt with by a system of negative values of IFLAG. To use Adams methods efficiently, we must let the code step along using the largest steps that will yield the requested accuracy everywhere. Between mesh points the answers are determined by interpolation very cheaply. If the step sizes chosen by the code are smaller than the interval of integration, there is no difficulty. But if the user wants a lot of output, the interval of integration may be substantially smaller than the step size the code could use. To handle this possibility efficiently the code must be allowed to integrate past TOUT internally and then compute the answers at TOUT by interpolation. The farther it is permitted to go, the more robust the code is with respect to a lot of output. For nearly all problems it is permissible to integrate past TOUT, though obviously some limit must be imposed. Our experience has been that users rarely ask for output at a spacing less than one tenth the natural step size, though it is quite common for them to want it more often than at mesh points alone. To deal with this DE permits the integration to go past TOUT internally, though never past T + 10*(TOUT − T). For nearly all problems this design means that the cost is insensitive to the number and placement of the TOUT and the user need give the matter no thought at all.

For some problems it is not permissible for DE to go past TOUT internally, e.g., the solution or its derivative is not defined beyond TOUT, or there is a discontinuity there. To warn the code of this, and so to have DE stop internally at TOUT, set IFLAG negative. Thus to initialize the code and have it stop at TOUT internally, use IFLAG = −1. When continuing an integration which is to be stopped internally at TOUT, use IFLAG = −2. Since the code is designed to return to the user in the event of trouble with a diagnostic in IFLAG, and be set to continue with a simple call again, there are negative values of IFLAG corresponding to the returns of IFLAG = 3, 4, 5. Thus if DE does not reach TOUT and returns with, for example, IFLAG = −3, then on calling DE again it will try once more to reach TOUT and stop internally at that point.

IFLAG = −1, −2, −3, −4, −5—same meaning as for the positive
values but the integration will terminate internally at
TOUT.

One should not use this feature to make up a table of the solution. The
only legitimate situation in which one would want to stop at TOUT and
then to continue the integration is in dealing with a discontinuity, or the
like. Such a situation ought to be followed by a restart. If the user fails to
do so, DE assumes this is an oversight and automatically restarts itself. To
prevent this situation arising accidentally, DE always returns with
IFLAG = +2 after a successful integration. Thus to stop internally at
TOUT on a continuation will require positive action by the user. This
protection is built into the code because abuse of the feature can be very
expensive and can adversely affect the accuracy.

To illustrate these points and how easy it is to use DE, we solve

$$y'_1 = y_2 y_3,$$
$$y'_2 = -y_1 y_3, \tag{3}$$
$$y'_3 = -0.51 y_1 y_2,$$

$$y_1(0) = 0.0, \qquad y_2(0) = 1.0, \qquad y_3(0) = 1.0. \tag{4}$$

The solutions of equations (3) and (4) are the Jacobian elliptic functions
$y_1(x) = \mathrm{sn}(x|0.51)$, $y_2(x) = \mathrm{cn}(x|0.51)$, $y_3(x) = \mathrm{dn}(x|0.51)$. They are periodic
with a quarter period of $K \doteq 1.862640802332739$. The true solutions at
$x = jK$ are obtained by using the table, along with use of the identity
expressing the periodicity, $y_i((j + 4)K) = y_i(jK)$ for $i = 1, 2, 3$ and all j.

j	$y_1(jK)$	$y_2(jK)$	$y_3(jK)$
0	0	1	1
1	1	0	0.7
2	0	−1	1
3	−1	0	0.7

We compute the solutions over one complete period and report the results
at quarter periods. Because the solutions oscillate between 1 and −1, the use
of relative error is inappropriate so we choose pure absolute error and
limit the local error to 10^{-5}. We also provide for continuing the integration
if DE cannot complete the quarter period in one call, although there is no
difficulty for this problem and error tolerance. Note that IFLAG is set to 1

to indicate the first call and that the integration can go past TOUT. This is set outside the loop so subsequent calls are made with the output value of IFLAG from the preceding call. The computation was done on a PDP-10 in a time sharing mode with output on a teletypewriter.

```
          EXTERNAL F
          DIMENSION Y(3)
          T = 0.0
          Y(1) = 0.0
          Y(2) = 1.0
          Y(3) = 1.0
          RELERR = 0.0
          ABSERR = 1.0E-5
          IFLAG = 1
          TOUT = 0.0
          DO 10 I = 1,4
          TOUT = TOUT + 1.8626408
1         CALL DE(F,3,Y,T,TOUT,RELERR,ABSERR,IFLAG)
          IF(IFLAG .EQ. 6) STOP
          IF(IFLAG .NE. 2) GO TO 1
          TYPE 100,T,(Y(J),J=1,3),IFLAG
100       FORMAT(4F15.7,I5)
10        CONTINUE
          STOP
          END
          SUBROUTINE F(X,Y,YP)
          DIMENSION Y(3),YP(3)
          YP(1) = Y(2)◆Y(3)
          YP(2) = -Y(1)◆Y(3)
          YP(3) = -0.51◆Y(1)◆Y(2)
          RETURN
          END
```

1.8626408	1.0000010	-0.0000005	0.6999998	2
3.7252816	0.0000018	-1.0000005	0.9999997	2
5.5879224	-1.0000018	-0.0000002	0.6999976	2
7.4505632	0.0000042	0.9999965	0.9999964	2

For those problems where it can integrate beyond the output point, DE is not affected very much by the number of print points requested. As a rule of thumb, increasing the number of print points by a factor up to 10 will cost very little. To illustrate this, we integrate equations (3) and (4) as above with 4, 40, 400, and 4000 print points in the interval of one period. A counter is included in the subroutine F to record the number of times the derivative functions are evaluated. We report only the solutions at the last print point and number of function evaluations in each case. Since 107 evaluations corresponds to about 50 steps we see that with 40 print points we are printing at about the average step size. Increasing to 400 print points is handled by interpolation and costs almost nothing. At 4000 print points we are limiting the step size because the code cannot go past a print point by more than 10 times the length of the interval of integration. If someone really wanted this much output, he should use STEP and INTRP which would give *any* number of print points with about 111 function evaluations.

```
         EXTERNAL F
         COMMON NEVAL
         DIMENSION Y(3)
         NO = 4
         DO 20 J = 1,4
         T = 0.0
         Y(1) = 0.0
         Y(2) = 1.0
         Y(3) = 1.0
         RELERR = 0.0
         ABSERR = 1.0E-5
         IFLAG = 1
         NEVAL = 0
         DELT = 4.0*1.8626408
         DO 10 I = 1,NO
         TOUT = DELT*FLOAT(I)/FLOAT(NO)
1        CALL DE(F,3,Y,T,TOUT,RELERR,ABSERR,IFLAG)
         IF(IFLAG .EQ. 6) STOP
         IF(IFLAG .NE. 2) GO TO 1
10       CONTINUE
         TYPE 99,NO
99       FORMAT(I15)
         TYPE 100,T,(Y(L),L=1,3),NEVAL
100      FORMAT(4F15.7,I5/)
20       NO = NO*10
         STOP
         END
         SUBROUTINE F(X,Y,YP)
         COMMON NEVAL
         DIMENSION Y(3),YP(3)
         NEVAL = NEVAL + 1
         YP(1) = Y(2)*Y(3)
         YP(2) = -Y(1)*Y(3)
         YP(3) = -0.51*Y(1)*Y(2)
         RETURN
         END
```

4				
7.4505632	0.0000042	0.9999965	0.9999964	107
40				
7.4505633	0.0000042	0.9999965	0.9999964	107
400				
7.4505632	-0.0000017	1.0000009	0.9999998	111
4000				
7.4505632	0.0000008	1.0000001	1.0000000	827

EXERCISE 1

DE is commonly used as follows:

$$\vdots$$

```
1     CALL  DE(F,NEQN,Y,T,TOUT,RELERR,ABSERR,IFLAG)
      IF(IABS(IFLAG).EQ.3) GO TO 1
```

$$\vdots$$

We call this the "do the best you can" mode; explain why.

EXERCISE 2

The equation $y' = -100y$ is usually described as stiff, but when solved on a PDP-10 with $y(0) = 10^{-4}$ and a pure absolute error tolerance of 10^{-5}, the following results were obtained:

On the interval $[0, 5]$ the code reached 5 and returned with IFLAG = 2 at a cost of 605 function evaluations. On the interval $[0, 10]$ the code returned at 8.854 with IFLAG = 5 at a cost of 1084 function evaluations.

Explain why the code does not report stiffness when it integrates the equation from 0 to 5.

EXERCISE 3

The solution of the following initial value problem for the ideal relay equation has jump discontinuities in its second derivative at multiples of $\pi/2$:

$$y'' + y + \text{sgn}(y) + 3 \sin 2x = 0,$$
$$y(0) = 0, \quad y'(0) = 3$$

where
$$\text{sgn}(y) = +1 \quad \text{if } y \geq 0,$$
$$= -1 \quad \text{if } y < 0.$$

The equation is not stiff, but when it was integrated with DE on a CDC 6600 over the interval $[0, 20]$ with a pure absolute error tolerance of 10^{-9}, IFLAG = 5 was returned. Explain why a false indication of stiffness was given.

EXERCISE 4

Suppose you wish to solve $\mathbf{y}' = \mathbf{f}(x, \mathbf{y})$, $\mathbf{y}(a)$ given on $[a, b]$, and then to solve $\mathbf{z}' = \mathbf{g}(x, \mathbf{z})$, $\mathbf{z}(b) = \mathbf{y}(b)$ on $[b, c]$, using DE. Explain why, if you initialize with IFLAG $= 1$ and use DE to integrate from a to b, you *must* restart at b to solve the other problem. Explain why, if you initialize with IFLAG $= -1$ and integrate from a to b, it is not necessary for you to restart at b to solve the other problem. Explain why, if you program F(X, Y, YP) to evaluate $\mathbf{f}(x, \mathbf{y})$ when $a \le x \le b$, and to evaluate $\mathbf{g}(x, \mathbf{y})$ when $b < x \le c$, you can initialize with IFLAG $= 1$ and integrate all the way from a to c. (The preferred way to solve this problem is to stop the code internally at b and to restart for the second integration.)

EXERCISE 5

Integrating $y' = \sin x$, $y(0) = 0$ from $x = 0$ to $x = \pi$ with DE illustrates the hazards of blindly depending on a code. DE will return with the value 0 as approximating $y(\pi) = 1$. However, if DE were asked to integrate to any $z < \pi$ and then from z to π, a good answer would be obtained. Can you explain what is happening? (Hint: the equation is evaluated only three times in the integration from 0 to π.)

STEP: Subroutine STEP solves a system of first order ordinary differential equations

$$\mathbf{y}'(x) = \mathbf{f}(x, \mathbf{y}(x)),$$

$$\mathbf{y}(a) = \mathbf{y}_0$$

using Adams methods. We anticipate most users will use this routine indirectly through the driver DE. There are problems, however, for which the driver is inadequate: for instance, when an error criterion not available in DE is desired, or when special action is to be taken at some unknown point in $[a, b]$. In these cases users must write their own drivers for STEP and INTRP. Before discussing the general form of such drivers, we caution the user that STEP does not check for invalid input parameters except in the case of too small a step size or error tolerance. Incorrect input can cause inefficiency, anomalous results, and even a complete breakdown.

The code STEP advances the solution of the differential equation one step and returns control to the calling program. Because of this and because of the information returned, it allows the integration to be monitored closely. It also allows great flexibility in the way the local error is controlled.

The subroutine is referenced

$$\text{CALL STEP(X,Y,F,NEQN,H,EPS,WT,START,}$$
$$1 \quad \text{HOLD,K,KOLD,CRASH,PHI,P,YP,PSI)}.$$

There are NEQN equations to be integrated and F is the name of a sub-routine of the form F(X,Y,YP) which evaluates the derivatives YP(L) = $y'_L(X)$ for L = 1,..., NEQN. The user must declare F in an EXTERNAL statement, declare START and CRASH in a LOGICAL statement, and dimension the arrays in the call list. The arrays must have dimensions of at least

$$\text{Y(NEQN),WT(NEQN),PHI(NEQN,16),P(NEQN),YP(NEQN),PSI(12)}.$$

Many of the arguments in the call list are solely for communication with INTRP so they are of no concern to the user. As with DE, the output of one call to STEP is nearly all the input for the next call so that STEP is only a little more trouble to use than DE.

Since STEP advances the solution one step at a time, it must be called repeatedly to integrate to a specified output point. And so, the typical situation is that a successful step has just been completed and the user wishes to take another step. On return from a successful step, X represents how far the integration has progressed, Y is the vector of solution components at X, YP is the vector of first derivatives of the solution at X, H is estimated by the code to be the largest step the code can take and still pass the error test, START = .FALSE., and CRASH = .FALSE.. The parameter EPS is the local error tolerance and the vector WT specifies the error criterion; we shall go into them later. The user may change EPS and WT as he wishes. For some kinds of error control it is necessary to change WT at each step. The user may alter H but ordinarily he should not. The code does a very good job of selecting H and the user should not override the choice without a good reason. It is most efficient to step past any desired output point and to use INTRP to obtain the solution and its derivative there. Sometimes it is not possible to integrate beyond a given point, as the equation may have a discontinuity or be undefined there, and it is necessary for the user to alter H so as to land at the desired point. The code will attempt to step to X + H when it is called. If this is not possible, it will reduce the step size and continue trying to complete a step until it either succeeds or realizes that the task is impossible. Thus while the user does not know how far the code will advance, he does know that it will not go beyond X + H. By manipulating H the user can arrange to stop at a given point. We emphasize that it is much more efficient and more accurate to step past output points and then use INTRP when this is possible. In summary, to continue after a successful step, the only quantity the user commonly changes before calling the code again is WT.

On the very first step the code must be initialized. This is done simply by supplying the initial values $X = a$, $Y(L) = y_L(a)$ for $L = 1, \ldots, NEQN$, and setting START $= .TRUE..$. The code chooses its own starting step size. The user must provide an initial value of H, but it serves only to show the direction of integration and to provide an upper bound on the length of the first step. Of course EPS and WT must be specified, as for any other step. Sometimes it is necessary to restart the code. The most common reasons are a change of direction, or because the derivatives are discontinuous and it is necessary to start afresh after defining them properly. This is handled like a first step but with the difference that some of the parameters are already set.

An error return occurs only when the code believes it impossible to take a step at all and meet the error request on the particular machine being used. The code returns with CRASH $= .TRUE.$, EPS increased to a new value which the code believes possible, and H set to an appropriate value. All other parameters are reset to the values that they had when STEP was called. If the user is willing to continue with the larger tolerance, he merely needs to call the code again without either restarting, or resetting CRASH.

On each step the code attempts to control the vector of local errors so that

$$\left(\sum_{L=1}^{NEQN} \left(\frac{\text{local error}_L}{WT(L)} \right)^2 \right)^{\frac{1}{2}} \leq EPS,$$

hence the estimated local error in each component satisfies

$$|\text{local error}_L/WT(L)| \leq EPS.$$

By choosing appropriate positive values of WT(L) a great variety of error criteria can be specified for individual components. For example if WT(L) = 1.0, the test is

$$|\text{local error}_L| \leq EPS.$$

Hence, a pure absolute error of no more than EPS is being asked for in the Lth component. If we define $WT(L) = |Y(L)|$, we shall have pure relative error; if we define WT(L) = |YP(L)|, we shall have error relative to the first derivative. The mixed relative-absolute error criterion of DE uses WT as follows. Let RE denote the desired relative error and AE the desired absolute error. The error tolerance EPS passed to STEP is EPS = max(RE,AE) and for each component, at each step, WT is defined as

$$WT(L) = |Y(L)|*RE/EPS + AE/EPS.$$

Obviously this corresponds to

$$|\text{local error}_L| \le |Y(L)|*RE + AE,$$

for each component. Another safe possibility is to take

$$WT(L) = \max(WT(L), |Y(L)|)$$

at each step. With WT(L) initialized to $|y_L(a)|$ this specifies relative error for increasing components and absolute error for decreasing components.

All components of WT must be non-zero or else we will probably encounter a divide check or an overflow. In particular, for the last criterion mentioned WT(L) cannot be initialized in the way described if the solution has a zero initial value. A safer way to proceed is to initialize WT(L) = 1.0, and this is quite commonly done. It leads to a reasonable error criterion which is always acceptable to the code.

INTRP: Subroutine INTRP(X,Y,XOUT,YOUT,YPOUT,NEQN,KOLD, PHI,PSI) calculates the solution vector YOUT and the first derivative vector YPOUT at any specified output point XOUT without impairing the step size selection of STEP. Results are most accurate if the user integrates the problem with STEP until just beyond XOUT, i.e., |X-HOLD| < |XOUT| ≤ |X|, and then interpolates. All input to INTRP is passed from STEP and is returned unaltered so the integration may continue as before. The user must supply storage for YOUT(NEQN) and YPOUT(NEQN) in the driving program, and supply the output point XOUT.

As an example, we integrate

$$y' = -y,$$

$$y(0) = 1$$

until the solution is less than 10^{-5}. Results are reported at intervals of one unit. Local error per step is limited to 10^{-6} relative to the solution.

If STEP crashes, the integration is terminated. The computation was done on a PDP-10.

```
         DIMENSION Y(1),WT(1),PHI(1,16),P(1),YP(1),PSI(12),
      1  YOUT(1),YPOUT(1)
         EXTERNAL F
         LOGICAL START,CRASH
         X = 0.0
         Y(1) = 1.0
         H = 0.1
         EPS = 1.0E-6
         WT(1) = ABS(Y(1))
         START = .TRUE.
         XOUT = 1.0
1        CALL STEP(X,Y,F,1,H,EPS,WT,START,
      1  HOLD,K,KOLD,CRASH,PHI,P,YP,PSI)
         IF(CRASH) STOP
         WT(1) = ABS(Y(1))
         IF(Y(1) .LT. 1.0E-5) GO TO 2
         IF(X .LT. XOUT) GO TO 1
         CALL INTRP(X,Y,XOUT,YOUT,YPOUT,1,KOLD,PHI,PSI)
         TYPE 100,XOUT,YOUT(1),YPOUT(1)
100      FORMAT(F15.2,2(1PE17.7))
         XOUT = XOUT + 1.0
         GO TO 1
2        TYPE 100,X,Y(1)
         STOP
         END
         SUBROUTINE F(X,Y,YP)
         DIMENSION Y(1),YP(1)
         YP(1) = -Y(1)
         RETURN
         END
```

```
    1.00     3.6787952E-01     -3.6787955E-01
    2.00     1.3533536E-01     -1.3533539E-01
    3.00     4.9787152E-02     -4.9787158E-02
    4.00     1.8315680E-02     -1.8315681E-02
    5.00     6.7379625E-03     -6.7379618E-03
    6.00     2.4787582E-03     -2.4787551E-03
    7.00     9.1188416E-04     -9.1188439E-04
    8.00     3.3546346E-04     -3.3546400E-04
    9.00     1.2341012E-04     -1.2341018E-04
   10.00     4.5400029E-05     -4.5400044E-05
   11.00     1.6701731E-05     -1.6701729E-05
   11.69     8.4095025E-06
```

In Chapter 12 we discuss other examples of computations requiring STEP rather than DE.

EXPLANATION OF THE CODES—STEP:

The code STEP uses a variable order, completely variable step version of the Adams formulas in a PECE combination with local extrapolation (i.e., a predictor of order k and a corrector of order $k + 1$) to solve the system of first order ordinary differential equations

$$\mathbf{y}'(x) = \mathbf{f}(x, \mathbf{y}(x)),$$

$$\mathbf{y}(a) = \mathbf{y}_0.$$

Each call to STEP normally advances the solution one step, that is, from x_n to $x_{n+1} = x_n + h$. Local error is controlled according to a criterion of a generalized error per unit step by varying the order and step size. Typically, changes of order are limited to one and changes of step size to halving and doubling. The order is limited to 12, which with extrapolation is effectively 13. The code is self-starting, requiring only that the initial conditions a and $\mathbf{y}_0$ be supplied in order to begin the integration at order 1.

The subroutine is referenced

$$\text{CALL STEP}(X,Y,F,NEQN,H,EPS,WT,START,$$
$$1 \quad \text{HOLD,K,KOLD,CRASH,PHI,P,YP,PSI}).$$

On input the parameters are:

X	—the independent variable set to x_n
Y(NEQN)	—the solution vector $\mathbf{y}_n$ at x_n
F	—externally declared subroutine of the form

$$\text{SUBROUTINE F}(X,Y,YP)$$

	to evaluate derivatives
NEQN	—number of first order equations to be integrated
H	—optimal step size for next step. Normally determined by code
EPS	—local error tolerance
WT(NEQN)	—weights for error control
START	—logical variable set .TRUE. for first step, .FALSE. otherwise
HOLD	—step size used in last successful step
K	—optimal order determined by code for next step
KOLD	—order used for last successful step

CRASH   —logical variable set .TRUE. if a step cannot be made, .FALSE. otherwise

PHI(NEQN,16)—array of modified divided differences. Columns 15 and 16 are used for propagated roundoff control

P(NEQN)   —predicted solution at x_n

YP(NEQN)   —derivative computed with corrected solution at x_n

PSI(12)   —coefficients $\psi_i(n) = h_n + h_{n-1} + \cdots + h_{n-i+1}$.

Upon return from a successful step the parameters are updated to correspond to x_{n+1}, and H and K are the "optimal" step size and order for continuing.

The code returns without taking a step (CRASH = .TRUE.) when the requested error tolerance is too small for the machine precision or the step size is less than four units of roundoff relative to the magnitude of the independent variable. STEP restores information as on input and estimates an appropriate step size and error tolerance before returning.

STEP is a straightforward implementation of the formulas in the preceding chapters; even the variable names correspond to earlier notation. We have been assuming only one equation, but extension to a system is obvious in all but one respect: simply apply the formula to each equation in the system. The exception is the local error estimate. With a system of equations there is a vector of local errors with components "local error$_L$". By using a weight vector WT with components WT(L) $\neq$ 0 we can specify different error measures for the various solution components. The code STEP requires that

$$\text{ERR} = \left(\sum_{L=1}^{\text{NEQN}} \left(\frac{\text{local error}_L}{\text{WT(L)}} \right)^2 \right)^{\frac{1}{2}} \leq \text{EPS}$$

for a successful step. This implies that

$$\left| \text{local error}_L / \text{WT(L)} \right| \leq \text{EPS}$$

for each component. The user can change EPS and WT at each step to impose whatever error control is appropriate; several kinds are discussed in the section on using the code. Similar measures of local error vectors for other orders and constant step size are computed and used to select the order and step size for the next step.

There are a great many references to the PHI array. It is stored by columns and always accessed that way in order to do efficient indexing in FORTRAN. The nested DO loops terminating at 230 and 435 for prediction and correction account for much of the overhead of the program, especially the prediction loop. Coding these loops in machine language might well increase the efficiency of the code substantially. We have removed from DO loops, and stored in temporary variables, those array elements which are accessed repeatedly in the loop. An optimizing compiler would do this automatically but others, like WATFOR, check indices each time so this can be a significant saving.

The code is divided into five logical units or blocks. With the exception of error returns, each block is entered through its first statement and left through its last statement. Within each block we have attempted to keep the flow of computation as linear as possible, subject to the constraints that the codes be written in FORTRAN and that they be efficient. All transfers of control have been particularly scrutinized with respect to the readability of the code. Temporary variables within a block are all named TEMP1, TEMP2, ... and no temporary variable is used for more than one purpose. The values of temporary variables are local to the block in which they appear; no temporary variable is used to pass information between blocks.

There is a prologue of comments explaining how to use the code. Each block is set off and has its own prologue that briefly explains what is done in the block. Within each block there are comments that explain what is being computed. Indenting conventions are used to enhance the readability of the code.

Block 0 first determines if the step size is sufficiently large to continue, i.e., $|H| \geq FOURU*|X|$ and, if it is, whether the error tolerance is feasible for the problem. These tests are explained in Chapter 9. If the input values are not acceptable, STEP "crashes" and returns with larger ones. Typically the computation proceeds and, if this is a first step, derivatives at the initial point are evaluated, a starting step size estimated as described in Chapter 7, and the need for the propagated roundoff control is determined. Usually the computation is not a first step and the code skips this initialization.

The coefficients of the formulas for prediction, correction, and so forth are computed in Block 1 exactly as described in Chapter 5. The coefficients $\sigma_i(n + 1)$ for converting the error estimates to constant step size are computed as described in Chapter 7. Some or all of these computations are omitted depending on the number of preceding steps taken with the same step size.

Block 2 does the prediction and error estimation. The prediction follows the description of Chapter 5; the alternative DO loop for the propagated roundoff control is explained in Chapter 9. The weighted local error ERR at order k is estimated, as well as the weighted local errors at constant

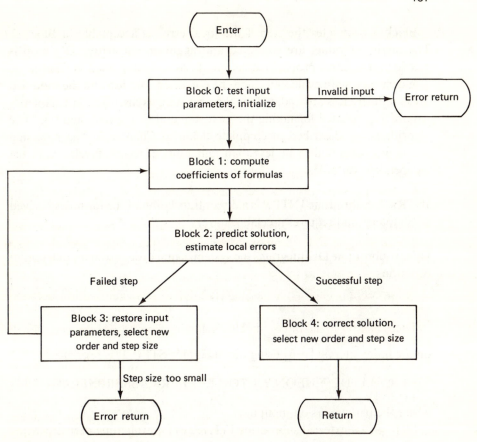

Block flow chart for STEP

step size for orders k, $k - 1$, $k - 2$ (ERK,ERKM1,ERKM2 respectively). The step is accepted if ERR $\leq$ EPS; otherwise it is rejected and control goes to Block 3. The possibility of lowering the order is considered. The reasoning behind these actions is explained in Chapter 7.

Block 3 restores information altered in the prediction when a step is unsuccessful and determines how to proceed in repeating the step. If it is the first or second failure, the step size is halved. If it is the third failure, the step size is halved and the order lowered to one, effectively a restart. If the step fails more than three times, the code estimates an "optimal" step size, thus allowing a faster and more precise adjustment of the step size. In no case is the new step size permitted to be less than four units of roundoff relative to the independent variable.

Block 4 completes the step if the local error is acceptable in Block 2. The predicted values are corrected and again an alternative DO loop is provided to control propagated roundoff errors as described in Chapter 9. The derivatives are evaluated using the corrected solution and the modified divided differences are updated for the next step, as discussed in Chapter 5. An order is selected for taking the next step and then a new step size. The algorithms are described in complete detail in Chapter 7. The new step size is not permitted to be less than four units of roundoff relative to the independent variable.

INTRP: Subroutine INTRP is a direct translation of the formulas derived in Chapter 5 and requires no elaboration.

DE: Subroutine DE integrates the system of first order ordinary differential equations

$$\mathbf{y}'(t) = \mathbf{f}(t, \mathbf{y}(t)),$$

$$\mathbf{y}(a) = \mathbf{y}_0$$

on the finite interval $[a, b]$ using the integrator STEP. It is referenced

CALL DE(F,NEQN,Y,T,TOUT,RELERR,ABSERR,IFLAG)

where the parameters on input are:
F   —external subroutine F(T,Y,YP) for evaluating derivatives
NEQN   —number of equations to be integrated
Y(NEQN)—solution vector $\mathbf{y}_0$
T   —independent variable set to initial value a
TOUT   —output point b
RELERR —relative local error tolerance
ABSERR —absolute local error tolerance
IFLAG   —indicator set
           ± 1—first step of integration.
The indicator IFLAG should be set negative only when the integration cannot proceed beyond TOUT; for instance, if the solution is not defined beyond TOUT or has a discontinuity there.

Normal return has IFLAG = 2, indicating that the integration has reached TOUT. Four conditions prevent the integration from reaching TOUT; these are signalled by IFLAG $\neq$ 2 on return. The first, IFLAG = ± 3, indicates an error request which appears to be impossible on the machine being used. DE returns with larger values for the error tolerances so the user has only to call DE again. IFLAG = ± 4 indicates that too many steps are needed to complete the integration. The maximum is defined as MAXNUM in a DATA statement and the version of DE presented here has MAXNUM = 500. If the user is prepared to stand the cost of continuing, he has only to call DE again. The value IFLAG = ±5 is returned when the equations appear to be stiff. A special code for such problems would normally be more efficient. If the user wishes to continue with DE, he just calls it again. These flags are returned negative only when the input value is negative. IFLAG = 6 signals invalid input and the integration is not attempted. The user must supply valid arguments and try again.

All calls to DE proceed as follows. Assuming that the input parameters are valid, the code sets variables related to the length and endpoint of the integration, initializes counters, and manipulates the input error tolerances to define the weight vector associated with the mixed error criterion. The variable TEND marks the end of the integration internal to DE. For input IFLAG positive, TEND = T + 10*(TOUT − T), so the step size for STEP is generally not impaired by the spacing of output points. For IFLAG negative, TEND = TOUT, which stops the integration internally at TOUT. If the call is a first call, or a situation treated as a first call, DE also sets the working quantities X and YY, initializes the step size, and stores the direction of integration. The integration proceeds with X and YY, not with the quantities T and Y passed to STEP.

The remainder of DE is executed repetitively until TOUT is reached or the code gives up. DE first checks if the work variable X is beyond TOUT, $|X - T| \geq |TOUT - T|$; if it is, DE interpolates the solution and returns. If DE cannot interpolate, it checks if IFLAG is negative and if the remaining interval of integration will require a step size too small for the machine precision. In such a case it extrapolates by Euler's method and returns. If the integration is not yet complete and fewer than MAXNUM steps have been taken, DE sets the weight vector for the local error test in STEP and attempts a step. If the step is successful, DE increments the two counters for the number of steps and the number of steps at low orders, and begins again the loop of checking to interpolate, extrapolate, etc. When STEP reports that it cannot advance because the error tolerances are too small, DE increases RELERR and ABSERR before returning. If MAXNUM steps have already been taken, DE tests for stiff equations, and then returns.

Although DE just supervises the use of STEP and INTRP, a sophisticated driver must consider a great many possible situations. The accompanying flow charts will clarify the possibilities and show how they are handled by DE.

184

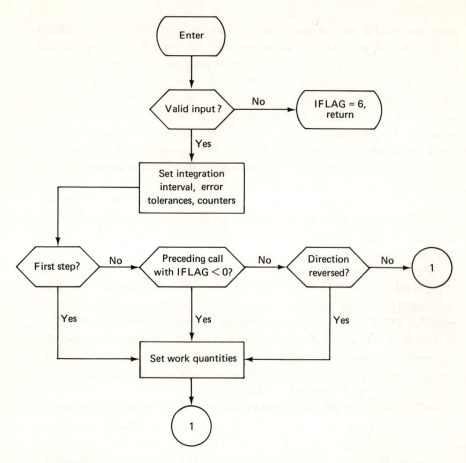

Flow chart for DE

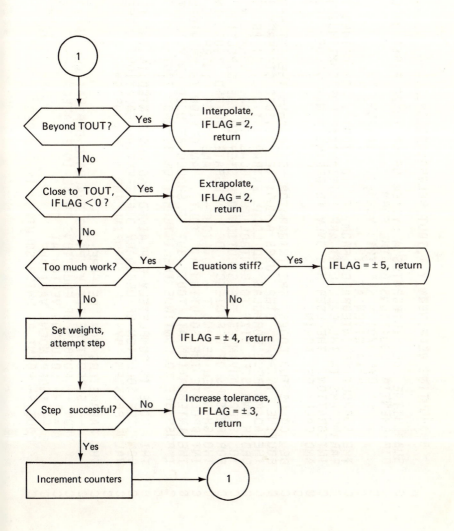

```
C     SUBROUTINE DE(F,NEQN,Y,T,TOUT,RELERR,ABSERR,IFLAG)          1
C                                                                 2
C     SUBROUTINE DE INTEGRATES A SYSTEM OF UP TO 20 FIRST ORDER ORDINARY   3
C     DIFFERENTIAL EQUATIONS OF THE FORM                          4
C             DY(I)/DT = F(T,Y(1),Y(2),...,Y(NEQN))               5
C             Y(I) GIVEN AT  T .                                  6
C     THE SUBROUTINE INTEGRATES FROM  T  TO  TOUT .  ON RETURN THE   7
C     PARAMETERS IN THE CALL LIST ARE INITIALIZED FOR CONTINUING THE   8
C     INTEGRATION.  THE USER HAS ONLY TO DEFINE A NEW VALUE TOUT   9
C     AND CALL  DE  AGAIN.                                        10
C                                                                11
C     DE  CALLS TWO CODES  THE INTEGRATOR STEP  AND THE INTERPOLATION   12
C     ROUTINE INTRP.  STEP  USES A MODIFIED DIVIDED DIFFERENCE FORM OF   13
C     THE ADAMS PECE FORMULAS AND LOCAL EXTRAPOLATION.  IT ADJUSTS THE   14
C     ORDER AND STEP SIZE TO CONTROL THE LOCAL ERROR.  NORMALLY EACH CALL   15
C     TO  STEP  ADVANCES THE SOLUTION ONE STEP IN THE DIRECTION OF TOUT .   16
C     FOR REASONS OF EFFICIENCY  DE  INTEGRATES BEYOND  TOUT  INTERNALLY,   17
C     THOUGH NEVER BEYOND  T + 10*(TOUT-T), AND CALLS  INTRP  TO   18
C     INTERPOLATE THE SOLUTION AT  TOUT .  AN OPTION IS PROVIDED TO STOP   19
C     THE INTEGRATION AT  TOUT  BUT IT SHOULD BE USED ONLY IF IT IS   20
C     IMPOSSIBLE TO CONTINUE THE INTEGRATION BEYOND  TOUT .        21
C                                                                22
C     THIS CODE IS COMPLETELY EXPLAINED AND DOCUMENTED IN THE TEXT   23
C     COMPUTER SOLUTION OF ORDINARY DIFFERENTIAL EQUATIONS:  THE INITIAL   24
C     VALUE PROBLEM BY L. F. SHAMPINE AND M. K. GORDON.           25
C                                                                26
C     THE PARAMETERS FOR  DE  ARE:                               27
C     F -- SUBROUTINE F(T,Y,YP) TO EVALUATE DERIVATIVES YP(I)=DY(I)/DT   28
C     NEQN -- NUMBER OF EQUATIONS TO BE INTEGRATED               29
C     Y(*) -- SOLUTION VECTOR AT  T                              30
C     T -- INDEPENDENT VARIABLE                                  31
C     TOUT -- POINT AT WHICH SOLUTION IS DESIRED                 32
```

```
C     RELERR,ABSERR -- RELATIVE AND ABSOLUTE ERROR TOLERANCES FOR LOCAL    33
C        ERROR TEST. AT EACH STEP THE CODE REQUIRES                        34
C        ABS(LOCAL ERROR) .LE. ABS(Y)*RELERR + ABSERR                      35
C        FOR EACH COMPONENT OF THE LOCAL ERROR AND SOLUTION VECTORS        36
C     IFLAG -- INDICATES STATUS OF INTEGRATION                             37
C                                                                          38
C                                                                          39
C   FIRST CALL TO DE --                                                    40
C                                                                          41
C   THE USER MUST PROVIDE STORAGE IN HIS CALLING PROGRAM FOR THE ARRAY     42
C   IN THE CALL LIST Y(NEQN), DECLARE F IN AN EXTERNAL STATEMENT,          43
C   SUPPLY THE SUBROUTINE F(T,Y,YP) TO EVALUATE                            44
C        DY(I)/DT = YP(I) = F(T,Y(1),Y(2),...,Y(NEQN))                     45
C   AND INITIALIZE THE PARAMETERS:                                         46
C     NEQN -- NUMBER OF EQUATIONS TO BE INTEGRATED                         47
C     Y(*) -- VECTOR OF INITIAL CONDITIONS                                 48
C     T -- STARTING POINT OF INTEGRATION                                   49
C     TOUT -- POINT AT WHICH SOLUTION IS DESIRED                           50
C     RELERR,ABSERR -- RELATIVE AND ABSOLUTE LOCAL ERROR TOLERANCES        51
C     IFLAG -- +1,-1. INDICATOR TO INITIALIZE THE CODE. NORMAL INPUT       52
C        IS +1. THE USER SHOULD SET IFLAG=-1 ONLY IF IT IS                 53
C        IMPOSSIBLE TO CONTINUE THE INTEGRATION BEYOND TOUT                54
C   ALL PARAMETERS EXCEPT F NEQN AND TOUT MAY BE ALTERED BY THE            55
C   CODE ON OUTPUT SO MUST BE VARIABLES IN THE CALLING PROGRAM.            56
C                                                                          57
C   OUTPUT FROM DE --                                                      58
C                                                                          59
C     NEQN -- UNCHANGED                                                    60
C     Y(*) -- SOLUTION AT T                                                61
C     T -- LAST POINT REACHED IN INTEGRATION. NORMAL RETURN HAS            62
C        T = TOUT                                                          63
C     TOUT -- UNCHANGED                                                    64
C     RELERR,ABSERR -- NORMAL RETURN HAS TOLERANCES UNCHANGED. IFLAG=3     65
C        SIGNALS TOLERANCES INCREASED                                      66
C     IFLAG = 2 -- NORMAL RETURN. INTEGRATION REACHED TOUT
```

```
C          = 3 -- INTEGRATION DID NOT REACH TOUT  BECAUSE ERROR             67
C               TOLERANCES TOO SMALL. RELERR , ABSERR INCREASED             68
C               APPROPRIATELY FOR CONTINUING                                69
C          = 4 -- INTEGRATION DID NOT REACH TOUT  BECAUSE MORE THAN         70
C               MAXNUM STEPS NEEDED                                         71
C          = 5 -- INTEGRATION DID NOT REACH TOUT  BECAUSE EQUATIONS         72
C               APPEAR TO BE STIFF                                          73
C          = 6 -- INVALID INPUT PARAMETERS (FATAL ERROR)                    74
C          THE VALUE OF IFLAG IS RETURNED NEGATIVE WHEN THE INPUT           75
C          VALUE IS NEGATIVE AND THE INTEGRATION DOES NOT REACH TOUT ,      76
C          I.E., -3, -4, -5.                                                77
C                                                                           78
C    SUBSEQUENT CALLS TO  DE --                                             79
C                                                                           80
C    SUBROUTINE DE  RETURNS WITH ALL INFORMATION NEEDED TO CONTINUE         81
C    THE INTEGRATION. IF THE INTEGRATION REACHED  TOUT  THE USER NEED       82
C    ONLY DEFINE A NEW  TOUT  AND CALL AGAIN. IF THE INTEGRATION DID NOT    83
C    REACH  TOUT  AND THE USER WANTS TO CONTINUE  HE JUST CALLS AGAIN.      84
C    THE OUTPUT VALUE OF  IFLAG  IS THE APPROPRIATE INPUT VALUE FOR         85
C    SUBSEQUENT CALLS.  THE ONLY SITUATION IN WHICH IT SHOULD BE ALTERED    86
C    IS TO STOP THE INTEGRATION INTERNALLY AT THE NEW  TOUT  I.E.           87
C    CHANGE OUTPUT  IFLAG=2  TO INPUT  IFLAG=-2 . ERROR TOLERANCES MAY      88
C    BE CHANGED BY THE USER BEFORE CONTINUING.  ALL OTHER PARAMETERS MUST   89
C    REMAIN UNCHANGED.                                                      90
C                                                                           91
C          LOGICAL START CRASH STIFF                                        92
C          DIMENSION Y(NEQN),PSI(12)                                        93
C          DIMENSION YY(20),WT(20),PHI(20,16),P(20),YP(20),YPOUT(20)        94
C          EXTERNAL F                                                       95
C************************************************************************   96
C* THE ONLY MACHINE DEPENDENT CONSTANT IS BASED ON THE MACHINE UNIT *       97
C* ROUNDOFF ERROR U WHICH IS THE SMALLEST POSITIVE NUMBER SUCH THAT *       98
C* 1.0+U .GT. 1.0 .  U MUST BE CALCULATED AND  FOURU=4.0*U INSERTED *       99
C************************************************************************  100
```

```
C*    IN THE FOLLOWING DATA STATEMENT BEFORE USING DE .  THE ROUTINE    *
C*    MACHIN CALCULATES U . FOURU AND TWOU=2.0*U MUST ALSO BE           *
C*    INSERTED IN SUBROUTINE . STEP BEFORE CALLING DE .                 *
      DATA FOURU /                                                      /
C***************************************************************************
C
C     THE CONSTANT  MAXNUM  IS THE MAXIMUM NUMBER OF STEPS ALLOWED IN ONE
C     CALL TO DE .  THE USER MAY CHANGE THIS LIMIT BY ALTERING THE
C     FOLLOWING STATEMENT
      DATA MAXNUM/500 /
C
C     THIS VERSION OF  DE  WILL HANDLE UP TO 20 EQUATIONS.  TO CHANGE THIS
C     NUMBER , ONLY THE NUMBER 20 IN THE DIMENSION STATEMENT ABOVE AND THE
C     NUMBER 20 IN THE FOLLOWING STATEMENT NEED BE CHANGED
C
C          ***                              ***
C
C     TEST FOR IMPROPER PARAMETERS
C
      IF(NEQN .LT. 1 .OR. NEQN .GT. 20) GO TO 10
      IF(T .EQ. TOUT) GO TO 10
      IF(RELERR .LT. 0.0 .OR. ABSERR .LT. 0.0) GO TO 10
      EPS = AMAX1(RELERR,ABSERR)
      IF(EPS .LE. 0.0) GO TO 10
      IF(IFLAG .EQ. 0) GO TO 10
      ISN = ISIGN(1,IFLAG)
      IFLAG = IABS(IFLAG)
      IF(IFLAG .EQ. 1) GO TO 20
      IF(T .NE. TOLD) GO TO 10
      IF(IFLAG .GE. 2 .AND. IFLAG .LE. 5) GO TO 20
   10 IFLAG = 6
      RETURN
C
C     ON EACH CALL SET INTERVAL OF INTEGRATION AND COUNTER FOR NUMBER OF
C     STEPS. ADJUST INPUT ERROR TOLERANCES TO DEFINE WEIGHT VECTOR FOR
C     SUBROUTINE STEP
```

101
102
103
104
105
106
107
108
109
110
111
112
113
114
115
116
117
118
119
120
121
122
123
124
125
126
127
128
129
130
131
132
133
134
135

```
 20  DEL = TOUT - T
     ABSDEL = ABS(DEL)
     TEND = T + 10.0*DEL
     IF(ISN .LT. 0) TEND = TOUT
     NOSTEP = 0
     KLE4 = 0
     STIFF = .FALSE.
     RELEPS = RELERR/EPS
     ABSEPS = ABSERR/EPS
     IF(IFLAG .EQ. 1) GO TO 30
     IF(ISNOLD .LT. 0) GO TO 30
     IF(DELSGN*DEL .GT. 0.0) GO TO 50

C
C    ON START AND RESTART ALSO SET WORK VARIABLES X AND YY(*), STORE THE
C    DIRECTION OF INTEGRATION AND INITIALIZE THE STEP SIZE
C
 30  START = .TRUE.
     X = T
     DO 40 L = 1, NEQN
 40    YY(L) = Y(L)
     DELSGN = SIGN(1.0, DEL)
     H = SIGN(AMAX1(ABS(TOUT-X), FOURU*ABS(X)), TOUT-X)

C
C    IF ALREADY PAST OUTPUT POINT, INTERPOLATE AND RETURN
C
 50  IF(ABS(X-T) .LT. ABSDEL) GO TO 60
     CALL INTRP(X, YY, TOUT, Y, YPOUT, NEQN, KOLD, PHI, PSI)
     IFLAG = 2
     T = TOUT
     TOLD = T
     ISNOLD = ISN
     RETURN

C
C    IF CANNOT GO PAST OUTPUT POINT AND SUFFICIENTLY CLOSE,
```

```
136
137
138
139
140
141
142
143
144
145
146
147
148
149
150
151
152
153
154
155
156
157
158
159
160
161
162
163
164
165
166
167
168
169
```

```
C     EXTRAPOLATE AND RETURN                                              170
C                                                                         171
   60 IF(ISN .GT. 0  .OR.   ABS(TOUT-X) .GE. FOURU*ABS(X)) GO TO 80       172
      H = TOUT - X                                                        173
      CALL F(X,YY,YP)                                                     174
      DO 70 L = 1,NEQN                                                    175
   70    Y(L) = YY(L) + H*YP(L)                                           176
      IFLAG = 2                                                           177
      T = TOUT                                                            178
      TOLD = T                                                            179
      ISNOLD = ISN                                                        180
      RETURN                                                              181
C                                                                         182
C     TEST FOR TOO MUCH WORK                                              183
C                                                                         184
   80 IF(NOSTEP .LT. MAXNUM) GO TO 100                                    185
      IFLAG = ISN*4                                                       186
      IF(STIFF) IFLAG = ISN*5                                             187
      DO 90 L = 1,NEQN                                                    188
   90    Y(L) = YY(L)                                                     189
      T = X                                                               190
      TOLD = T                                                            191
      ISNOLD = 1                                                          192
      RETURN                                                              193
C                                                                         194
C     LIMIT STEP SIZE, SET WEIGHT VECTOR AND TAKE A STEP                  195
C                                                                         196
  100 H = SIGN(AMIN1(ABS(H),ABS(TEND-X)),H)                               197
      DO 110 L = 1 NEQN                                                   198
  110    WT(L) = RE(EPS*ABS(YY(L)) + ABSEPS                               199
      CALL STEP(X,YY F NEQN H,EPS,WT START,                              200
     1 HOLD,K,KOLD,CRASH,PHI,P,YP,PSI)                                    201
C                                                                         202
C     TEST FOR TOLERANCES TOO SMALL                                      203
```

```
C
      IF(.NOT.CRASH) GO TO 130
      IFLAG = ISN*3
      RELERR = EPS*RELEPS
      ABSERR = EPS*ABSEPS
      DO 120 L = 1 NEQN
120   Y(L) = YY(L)
      T = X
      TOLD = T
      ISNOLD = 1
      RETURN
C
C     AUGMENT COUNTER ON WORK AND TEST FOR STIFFNESS
C
130   NOSTEP = NOSTEP + 1
      KLE4 = KLE4 + 1
      IF(KOLD .GT. 4) KLE4 = 0
      IF(KLE4 .GE. 50) STIFF = .TRUE.
      GO TO 50
      END
```

204
205
206
207
208
209
210
211
212
213
214
215
216
217
218
219
220
221
222
223

```
C         SUBROUTINE STEP(X,Y,F,NEQN,H,EPS,WT,START,                      1
C        1 HOLD,K,KOLD,CRASH,PHI,P,YP,PSI)                               2
C                                                                         3
C         SUBROUTINE  STEP  INTEGRATES A SYSTEM OF FIRST ORDER ORDINARY   4
C         DIFFERENTIAL EQUATIONS ONE STEP  NORMALLY FROM X TO X+H, USING A 5
C         MODIFIED DIVIDED DIFFERENCE FORM OF THE ADAMS PECE FORMULAS. LOCAL 6
C         EXTRAPOLATION IS USED TO IMPROVE ABSOLUTE STABILITY AND ACCURACY. 7
C         THE CODE ADJUSTS ITS ORDER AND STEP SIZE TO CONTROL THE LOCAL ERROR 8
C         PER UNIT STEP IN A GENERALIZED SENSE.  SPECIAL DEVICES ARE INCLUDED 9
C         TO CONTROL ROUNDOFF ERROR AND TO DETECT WHEN THE USER IS REQUESTING 10
C         TOO MUCH ACCURACY.                                            11
C                                                                        12
C         THIS CODE IS COMPLETELY EXPLAINED AND DOCUMENTED IN THE TEXT   13
C         COMPUTER SOLUTION OF ORDINARY DIFFERENTIAL EQUATIONS:  THE INITIAL 14
C         VALUE PROBLEM BY L. F. SHAMPINE AND M. K. GORDON.             15
C                                                                        16
C         THE PARAMETERS REPRESENT:                                     17
C           X -- INDEPENDENT VARIABLE                                   18
C           Y(*) -- SOLUTION VECTOR AT X                                19
C           YP(*) -- DERIVATIVE OF SOLUTION VECTOR AT  X  AFTER SUCCESSFUL 20
C                STEP                                                    21
C           NEQN -- NUMBER OF EQUATIONS TO BE INTEGRATED                22
C           H -- APPROPRIATE STEP SIZE FOR NEXT STEP.  NORMALLY DETERMINED BY 23
C                CODE                                                    24
C           EPS -- LOCAL ERROR TOLERANCE.  MUST BE VARIABLE             25
C           WT(*) -- VECTOR OF WEIGHTS FOR ERROR CRITERION              26
C           START -- LOGICAL VARIABLE SET .TRUE. FOR FIRST STEP,  .FALSE. 27
C                OTHERWISE                                               28
C           HOLD -- STEP SIZE USED FOR LAST SUCCESSFUL STEP             29
C           K -- APPROPRIATE ORDER FOR NEXT STEP (DETERMINED BY CODE)   30
C                                                                        31
```

```
C     KOLD -- ORDER USED FOR LAST SUCCESSFUL STEP
C     CRASH -- LOGICAL VARIABLE SET .TRUE. WHEN NO STEP CAN BE TAKEN,
C              .FALSE. OTHERWISE.
C     THE ARRAYS PHI PSI ARE REQUIRED FOR THE INTERPOLATION SUBROUTINE
C     INTRP . THE ARRAY P IS INTERNAL TO THE CODE.
C
C     INPUT TO STEP
C
C        FIRST CALL --
C
C     THE USER MUST PROVIDE STORAGE IN HIS DRIVER PROGRAM FOR ALL ARRAYS
C     IN THE CALL LIST, NAMELY
C
C       DIMENSION Y(NEQN),WT(NEQN),PHI(NEQN,16),P(NEQN),YP(NEQN),PSI(12)
C
C     THE USER MUST ALSO DECLARE START AND CRASH LOGICAL VARIABLES
C     AND F AN EXTERNAL SUBROUTINE, SUPPLY THE SUBROUTINE F(X,Y,YP)
C     TO EVALUATE
C       DY(I)/DX = YP(I) = F(X,Y(1),Y(2),...,Y(NEQN))
C     AND INITIALIZE ONLY THE FOLLOWING PARAMETERS:
C       X -- INITIAL VALUE OF THE INDEPENDENT VARIABLE
C       Y(*) -- VECTOR OF INITIAL VALUES OF DEPENDENT VARIABLES
C       NEQN -- NUMBER OF EQUATIONS TO BE INTEGRATED
C       H -- NOMINAL STEP SIZE INDICATING DIRECTION OF INTEGRATION
C            AND MAXIMUM SIZE OF STEP. MUST BE VARIABLE
C       EPS -- LOCAL ERROR TOLERANCE PER STEP. MUST BE VARIABLE
C       WT(*) -- VECTOR OF NON-ZERO WEIGHTS FOR ERROR CRITERION
C       START -- .TRUE.
C
C     STEP REQUIRES THE L2 NORM OF THE VECTOR WITH COMPONENTS
C     LOCAL ERROR(L)/WT(L) BE LESS THAN EPS FOR A SUCCESSFUL STEP.   THE
C     ARRAY WT ALLOWS THE USER TO SPECIFY AN ERROR TEST APPROPRIATE
C     FOR HIS PROBLEM. FOR EXAMPLE,
```

```
C      WT(L) = 1.0  SPECIFIES ABSOLUTE ERROR,
C            = ABS(Y(L)) ERROR RELATIVE TO THE MOST RECENT VALUE OF THE
C              L-TH COMPONENT OF THE SOLUTION,
C            = ABS(YP(L)) ERROR RELATIVE TO THE MOST RECENT VALUE OF
C              THE L-TH COMPONENT OF THE DERIVATIVE.
C            = AMAX1(WT(L),ABS(Y(L))) ERROR RELATIVE TO THE LARGEST
C              MAGNITUDE OF L-TH COMPONENT OBTAINED SO FAR,
C            = ABS(Y(L))*RELERR/EPS + ABSERR/EPS SPECIFIES A 'MIXED
C              RELATIVE-ABSOLUTE TEST WHERE RELERR IS RELATIVE
C              ERROR, ABSERR IS ABSOLUTE ERROR AND EPS =
C              AMAX1(RELERR,ABSERR)
C
C      SUBSEQUENT CALLS --
C
C   SUBROUTINE STEP IS DESIGNED SO THAT ALL INFORMATION NEEDED TO
C   CONTINUE THE INTEGRATION INCLUDING THE STEP SIZE H AND THE ORDER
C   K IS RETURNED WITH EACH STEP. WITH THE EXCEPTION OF THE STEP
C   SIZE THE ERROR TOLERANCE AND THE WEIGHTS, NONE OF THE PARAMETERS
C   SHOULD BE ALTERED. THE ARRAY WT MUST BE, UPDATED AFTER EACH STEP
C   TO MAINTAIN RELATIVE ERROR TESTS LIKE THOSE ABOVE. NORMALLY THE
C   INTEGRATION IS CONTINUED JUST BEYOND THE DESIRED ENDPOINT AND THE
C   SOLUTION INTERPOLATED THERE WITH SUBROUTINE INTRP. IF IT IS
C   IMPOSSIBLE TO INTEGRATE BEYOND THE ENDPOINT THE STEP SIZE MAY BE
C   REDUCED TO HIT THE ENDPOINT SINCE THE CODE WILL NOT TAKE A STEP
C   LARGER THAN THE H INPUT. CHANGING THE DIRECTION OF INTEGRATION,
C   I.E. THE SIGN OF H, REQUIRES THE USER SET START = .TRUE. BEFORE
C   CALLING STEP AGAIN. THIS IS THE ONLY SITUATION IN WHICH START
C   SHOULD BE ALTERED.
C
C      OUTPUT FROM  STEP --
C
C      SUCCESSFUL STEP --
```

```
C      THE SUBROUTINE RETURNS AFTER EACH SUCCESSFUL STEP WITH START AND
C      CRASH SET .FALSE.  X REPRESENTS THE INDEPENDENT VARIABLE
C      ADVANCED ONE STEP OF LENGTH HOLD FROM ITS VALUE ON INPUT AND Y
C      THE SOLUTION VECTOR AT THE NEW VALUE OF X . ALL OTHER PARAMETERS
C      REPRESENT INFORMATION CORRESPONDING TO THE NEW X NEEDED TO
C      CONTINUE THE INTEGRATION.
C
C         UNSUCCESSFUL STEP --
C
C      WHEN THE ERROR TOLERANCE IS TOO SMALL FOR THE MACHINE PRECISION,
C      THE SUBROUTINE RETURNS WITHOUT TAKING A STEP AND CRASH = .TRUE.
C      AN APPROPRIATE STEP SIZE AND ERROR TOLERANCE FOR CONTINUING ARE
C      ESTIMATED AND ALL OTHER INFORMATION IS RESTORED AS UPON INPUT
C      BEFORE RETURNING. TO CONTINUE WITH THE LARGER TOLERANCE, THE USER
C      JUST CALLS THE CODE AGAIN. A RESTART IS NEITHER REQUIRED NOR
C      DESIRABLE.
C
C
      LOGICAL START,CRASH,PHASE1,NORND
      DIMENSION Y(NEQN),WT(NEQN),PHI(NEQN,16),P(NEQN),YP(NEQN),PSI(12)
      DIMENSION ALPHA(12),BETA(12),SIG(13),W(12),V(12),G(13),
     1 GSTR(13),TWO(13)
      EXTERNAL F
C***********************************************************************
C*  THE ONLY MACHINE DEPENDENT CONSTANTS ARE BASED ON THE MACHINE UNIT *
C*  ROUNDOFF ERROR  U  WHICH IS THE SMALLEST POSITIVE NUMBER SUCH THAT *
C*  1.0+U .GT. 1.0 .  THE USER MUST CALCULATE  U  AND INSERT           *
C*  TWOU=2.0*U  AND  FOURU=4.0*U  IN THE DATA STATEMENT BEFORE CALLING *
C*  THE CODE.  THE ROUTINE  MACHIN  CALCULATES  U .                    *
C      DATA TWOU,FOURU /
C***********************************************************************
      DATA TWO/2.0,4.0,8.0,16.0,32.0,64.0,128.0,256.0,512.0,1024.0,
     1 2048.0,4096.0,8192.0/
```

```
      DATA GSTR/0.500,0.0833,0.0417,0.0264,0.0188,0.0143,0.0114,0.00936,   130
     1 0.00789,0.00679,0.00592,0.00524,0.00468/                            131
      DATA G(1),G(2)/1.6,0.5/,SIG(1)/1.6/                                  132
                                                                           133
C           ***      BEGIN BLOCK 0      ***                                134
C     CHECK IF STEP SIZE OR ERROR TOLERANCE IS TOO SMALL FOR MACHINE       135
C     PRECISION. IF FIRST STEP, INITIALIZE PHI ARRAY AND ESTIMATE A        136
C     STARTING STEP SIZE.                                                  137
C           ***                                                            138
                                                                           139
C     IF STEP SIZE IS TOO SMALL, DETERMINE AN ACCEPTABLE ONE               140
                                                                           141
      CRASH = .TRUE.                                                       142
      IF(ABS(H) .GE. FOURU*ABS(X)) GO TO 5                                 143
      H = SIGN(FOURU*ABS(X),H)                                             144
      RETURN                                                               145
    5 P5EPS = 0.5*EPS                                                      146
                                                                           147
C     IF ERROR TOLERANCE IS TOO SMALL, INCREASE IT TO AN ACCEPTABLE VALUE  148
                                                                           149
      ROUND = 0.0                                                          150
      DO 10 L = 1,NEQN                                                     151
   10   ROUND = ROUND + (Y(L)/WT(L))**2                                    152
      ROUND = TWOU*SQRT(ROUND)                                             153
      IF(P5EPS .GE. ROUND) GO TO 15                                        154
      EPS = 2.0*ROUND*(1.0 + FOURU)                                        155
      RETURN                                                               156
   15 CRASH = .FALSE.                                                      157
      IF(.NOT.START) GO TO 99                                             158
                                                                           159
C     INITIALIZE.  COMPUTE APPROPRIATE STEP SIZE FOR FIRST STEP            160
C                                                                          161
      CALL F(X,Y,YP)                                                       162
                                                                           163
```

```
            SUM = 0.0                                                        164
            DO 20 L = 1,NEQN                                                 165
              PHI(L,1) = YP(L)                                               166
              PHI(L,2) = 0.0                                                 167
20          SUM = SUM + (YP(L)/WT(L))**2                                     168
          SUM = SQRT(SUM)                                                    169
          ABSH = ABS(H)                                                      170
          IF(EPS .LT. 16.0*SUM*H*H) ABSH = 0.25*SQRT(EPS/SUM)               171
          H = SIGN(AMAX1(ABSH,FOURU*ABS(X)),H)                              172
          HOLD = 0.0                                                        173
          K = 1                                                             174
          KOLD = 0                                                         175
          START = .FALSE.                                                   176
          PHASE1 = .TRUE.                                                   177
          NORND = .TRUE.                                                    178
          IF(P5EPS .GT. 100.0*ROUND) GO TO 99                              179
          NORND = .FALSE.                                                   180
          DO 25 L = 1,NEQN                                                  181
25          PHI(L,15) = 0.0                                                 182
99        IFAIL = 0                                                         183
C     ***        END BLOCK 0        ***                                     184
C                                                                           185
C     ***        BEGIN BLOCK 1        ***                                   186
C     COMPUTE COEFFICIENTS OF FORMULAS FOR THIS STEP.  AVOID COMPUTING      187
C     THOSE QUANTITIES NOT CHANGED WHEN STEP SIZE IS NOT CHANGED.           188
C     ***                                                                   189
C                                                                           190
100       KP1 = K+1                                                        191
          KP2 = K+2                                                        192
          KM1 = K-1                                                        193
          KM2 = K-2                                                        194
C                                                                           195
C     NS IS THE NUMBER OF STEPS TAKEN WITH SIZE H, INCLUDING THE CURRENT    196
C     ONE.  WHEN K.LT.NS, NO COEFFICIENTS CHANGE                            197
C                                                                           198
```

```
199       IF(H .NE. HOLD) NS = 0
200       NS = MIN0(NS+1,KOLD+1)
201       NSP1 = NS+1
202       IF (K .LT. NS) GO TO 199
203 C
204 C     COMPUTE THOSE COMPONENTS OF ALPHA(*),BETA(*),PSI(*),SIG(*) WHICH
205 C     ARE CHANGED
206 C
207       BETA(NS) = 1.0
208       REALNS = NS
209       ALPHA(NS) = 1.0/REALNS
210       TEMP1 = H*REALNS
211       SIG(NSP1) = 1.0
212       IF(K .LT. NSP1) GO TO 110
213       DO 105 I = NSP1,K
214         IM1 = I-1
215         TEMP2 = PSI(IM1)
216         PSI(IM1) = TEMP1
217         BETA(I) = BETA(IM1)*PSI(IM1)/TEMP2
218         TEMP1 = TEMP2 + H
219         ALPHA(I) = H/TEMP1
220         REALI = I
221 105     SIG(I+1) = REALI*ALPHA(I)*SIG(I)
222 110   PSI(K) = TEMP1
223 C
224 C     COMPUTE COEFFICIENTS G(*)
225 C
226 C     INITIALIZE V(*) AND SET W(*).   G(2) IS SET IN DATA STATEMENT
227 C
228       IF(NS .GT. 1) GO TO 120
229       DO 115 IQ = 1,K
230         TEMP3 = IQ*(IQ+1)
231         V(IQ) = 1.0/TEMP3
232 115     W(IQ) = V(IQ)
233       GO TO 140
```

```
C
C      IF ORDER WAS RAISED, UPDATE DIAGONAL PART OF V(*)
C
  120 IF(K .LE. KOLD) GO TO 130
      TEMP4 = K*KP1
      V(K) = 1.0/TEMP4
      NSM2 = NS-2
      IF(NSM2 .LT. 1) GO TO 130
      DO 125 J = 1, NSM2
        I = K-J
  125   V(I) = V(I) - ALPHA(J+1)*V(I+1)
C
C      UPDATE V(*) AND SET W(*)
C
  130 LIMIT1 = KP1 - NS
      TEMP5 = ALPHA(NS)
      DO 135 IQ = 1,LIMIT1
        V(IQ) = V(IQ) - TEMP5*V(IQ+1)
  135   W(IQ) = V(IQ)
      G(NSP1) = W(1)
C
C      COMPUTE THE G(*) IN THE WORK VECTOR W(*)
C
  140 NSP2 = NS + 2
      IF(KP1 .LT. NSP2) GO TO 199
      DO 150 I = NSP2,KP1
        LIMIT2 = KP2 - I
        TEMP6 = ALPHA(I-1)
        DO 145 IQ = 1,LIMIT2
  145     W(IQ) = W(IQ) - TEMP6*W(IQ+1)
  150   G(I) = W(1)
  199 CONTINUE
C
C ***         END BLOCK 1        ***
```

```
268   C        ***      BEGIN BLOCK 2      ***
269   C     PREDICT A SOLUTION P(*), EVALUATE DERIVATIVES USING PREDICTED
270   C     SOLUTION, ESTIMATE LOCAL ERROR AT ORDER K AND ERRORS AT ORDERS K,
271   C     K-1, K-2 AS IF CONSTANT STEP SIZE WERE USED.
272   C        ***
273
274   C     CHANGE PHI TO PHI STAR
275
276         IF(K .LT. NSP1) GO TO 215
277         DO 210 I = NSP1,K
278         TEMP1 = BETA(I)
279         DO 205 L = 1,NEQN
280   205   PHI(L,I) = TEMP1*PHI(L,I)
281   210   CONTINUE
282
283   C     PREDICT SOLUTION AND DIFFERENCES
284
285   215   DO 220 L = 1,NEQN
286         PHI(L,KP2) = PHI(L,KP1)
287         PHI(L,KP1) = 0.0
288   220   P(L) = 0.0
289         DO 230 J = 1,K
290         I = KP1 - J
291         IP1 = I+1
292         TEMP2 = G(I)
293         DO 225 L = 1,NEQN
294         P(L) = P(L) + TEMP2*PHI(L,I)
295   225   PHI(L,I) = PHI(L,I) + PHI(L,IP1)
296   230   CONTINUE
297         IF(NORND) GO TO 240
298         DO 235 L = 1,NEQN
299         TAU = H*P(L) - PHI(L,15)
300         P(L) = Y(L) + TAU
301   235   PHI(L,16) = (P(L) - Y(L)) - TAU
302         GO TO 250
```

```
240 DO 245 L = 1,NEQN
245   P(L) = Y(L) + H*P(L)
250 XOLD = X
    X = X + H
    ABSH = ABS(H)
    CALL F(X,P,YP)
C
C     ESTIMATE ERRORS AT ORDERS K,K-1,K-2
C
    ERKM2 = 0.0
    ERKM1 = 0.0
    ERK = 0.0
    DO 265 L = 1,NEQN
      TEMP3 = 1.0/WT(L)
      TEMP4 = YP(L) - PHI(L,1)
      IF(KM2)265,260,255
255   ERKM2 = ERKM2 + ((PHI(L,KM1)+TEMP4)*TEMP3)**2
260   ERKM1 = ERKM1 + ((PHI(L,K)+TEMP4)*TEMP3)**2
265   ERK = ERK + (TEMP4*TEMP3)**2
    IF(KM2)280,275,270
270   ERKM2 = ABSH*SIG(KM1)*GSTR(KM2)*SQRT(ERKM2)
275   ERKM1 = ABSH*SIG(K)*GSTR(KM1)*SQRT(ERKM1)
280   TEMP5 = ABSH*SQRT(ERK)
    ERR = TEMP5*(G(K)-G(KP1))
    ERK = TEMP5*SIG(KP1)*GSTR(K)
    KNEW = K
C
C     TEST IF ORDER SHOULD BE LOWERED
C
    IF(KM2)299,290,285
285 IF(AMAX1(ERKM1,ERKM2) .LE. ERK) KNEW = KM1
    GO TO 299
290 IF(ERKM1 .LE. 0.5*ERK) KNEW = KM1
C
```

303
304
305
306
307
308
309
310
311
312
313
314
315
316
317
318
319
320
321
322
323
324
325
326
327
328
329
330
331
332
333
334
335
336

```
C     TEST IF STEP SUCCESSFUL                                          337
C                                                                      338
299   IF(ERR .LE. EPS) GO TO 400                                       339
C     ***          END BLOCK 2      ***                                340
C                                                                      341
C     ***          BEGIN BLOCK 3    ***                                342
C     THE STEP IS UNSUCCESSFUL.   RESTORE X PHI(*,*), PSI(*)           343
C     IF THIRD CONSECUTIVE FAILURE SET ORDER TO ONE.  IF STEP FAILS MORE 344
C     THAN THREE TIMES, CONSIDER AN OPTIMAL STEP SIZE.  DOUBLE ERROR   345
C     TOLERANCE AND RETURN IF ESTIMATED STEP SIZE IS TOO SMALL FOR MACHINE 346
C     PRECISION.                                                       347
C                                                                      348
C     ***                                                              349
C     RESTORE X, PHI(*,*) AND PSI(*)                                   350
C                                                                      351
      PHASE1 = .FALSE.                                                 352
      X = XOLD                                                         353
      DO 310 I = 1,K                                                   354
        TEMP1 = 1.0/BETA(I)                                           355
        IP1 = I+1                                                      356
        DO 305 L = 1,NEQN                                              357
305       PHI(L,I) = TEMP1*(PHI(L,I) - PHI(L,IP1))                    358
310   CONTINUE                                                         359
      IF(K .LT. 2) GO TO 320                                           360
      DO 315 I = 2,K                                                   361
315     PSI(I-1) = PSI(I) - H                                         362
C                                                                      363
C     ON THIRD FAILURE, SET ORDER TO ONE.  THEREAFTER, USE OPTIMAL STEP 364
C     SIZE                                                             365
C                                                                      366
320   IFAIL = IFAIL + 1                                                367
      TEMP2 = 0.5                                                      368
      IF(IFAIL - 3) 335,330,325                                       369
325   IF(P5EPS .LT. 0.25*ERK) TEMP2 = SQRT(P5EPS/ERK)                 370
```

```
330 KNEW = 1                                                          371
335 H = TEMP2*H                                                       372
    K = KNEW                                                          373
    IF(ABS(H) .GE. FOURU*ABS(X)) GO TO 340                           374
    CRASH = .TRUE.                                                    375
    H = SIGN(FOURU*ABS(X),H)                                          376
    EPS = EPS + EPS                                                   377
    RETURN                                                            378
340 GO TO 100                                                         379
C                                                                     380
C***       END BLOCK 3       ***                                      381
C                                                                     382
C***       BEGIN BLOCK 4     ***                                      383
C   THE STEP IS SUCCESSFUL.  CORRECT THE PREDICTED SOLUTION, EVALUATE 384
C   THE DERIVATIVES USING THE CORRECTED SOLUTION AND UPDATE THE       385
C   DIFFERENCES. DETERMINE BEST ORDER AND STEP SIZE FOR NEXT STEP .   386
C                                                                     387
C***                         ***                                      388
C                                                                     389
400 KOLD = K                                                          390
    HOLD = H                                                          391
C                                                                     392
C   CORRECT AND EVALUATE                                              393
C                                                                     394
    TEMP1 = H*G(KP1)                                                  395
    IF(NORND) GO TO 410                                               396
    DO 405 L = 1 NEQN                                                 397
      RHO = TEMP1*(YP(L) - PHI(L,1)) - PHI(L,16)                      398
      Y(L) = P(L) + RHO                                               399
405   PHI(L,15) = (Y(L) - P(L)) - RHO                                 400
    GO TO 420                                                         401
410 DO 415 L = 1,NEQN                                                 402
415   Y(L) = P(L) + TEMP1*(YP(L) - PHI(L,1))                          403
420 CALL F(X,Y,YP)                                                    404
C
C   UPDATE DIFFERENCES FOR NEXT STEP
C
```

```
      DO 425 L = 1, NEQN
        PHI(L,KP1) = YP(L) - PHI(L,1)
425     PHI(L,KP2) = PHI(L,KP1) - PHI(L,KP2)
      DO 435 I = 1, K
        DO 430 L = 1, NEQN
430       PHI(L,I) = PHI(L,I) + PHI(L,KP1)
435   CONTINUE
C
C     ESTIMATE ERROR AT ORDER K+1 UNLESS:
C       IN FIRST PHASE WHEN ALWAYS RAISE ORDER,
C       ALREADY DECIDED TO LOWER ORDER
C       STEP SIZE NOT CONSTANT SO ESTIMATE UNRELIABLE
C
      ERKP1 = 0.0
      IF(KNEW .EQ. KM1 .OR. K .EQ. 12) PHASE1 = .FALSE.
      IF(PHASE1) GO TO 450
      IF(KNEW .EQ. KM1) GO TO 455
      IF(KP1 .GT. NS) GO TO 460
      DO 440 L = 1, NEQN
440   ERKP1 = ERKP1 + (PHI(L,KP2)/WT(L))**2
      ERKP1 = ABSH*GSTR(KP1)*SQRT(ERKP1)
C
C     USING ESTIMATED ERROR AT ORDER K+1, DETERMINE APPROPRIATE ORDER
C     FOR NEXT STEP
C
      IF(K .GT. 1) GO TO 445
      IF(ERKP1 .GE. 0.5*ERK) GO TO 460
      GO TO 450
445   IF(ERKM1 .LE. AMIN1(ERK, ERKP1)) GO TO 455
      IF(ERKP1 .GE. ERK .OR. K .EQ. 12) GO TO 460
C
C     HERE ERKP1 .LT. ERK .LT. AMAX1(ERKM1,ERKM2) ELSE ORDER WOULD HAVE
C     BEEN LOWERED IN BLOCK 2. THUS ORDER IS TO BE RAISED
C
```

```
C     RAISE ORDER
C
  450 K = KP1                                                      439
      ERK = ERKP1                                                  440
      GO TO 460                                                    441
                                                                   442
C                                                                  443
C     LOWER ORDER                                                  444
C                                                                  445
  455 K = KM1                                                      446
      ERK = ERKM1                                                  447
                                                                   448
C                                                                  449
C     WITH NEW ORDER DETERMINE APPROPRIATE STEP SIZE FOR NEXT STEP 450
C                                                                  451
  460 HNEW = H + H                                                 452
      IF(PHASE1) GO TO 465                                         453
      IF(P5EPS .GE. ERK*TWO(K+1)) GO TO 465                        454
      HNEW = H                                                     455
      IF(P5EPS .GE. ERK) GO TO 465                                 456
      TEMP2 = K+1                                                  457
      R = (P5EPS/ERK)**(1.0/TEMP2)                                 458
      HNEW = ABSH*AMAX1(0.5,AMIN1(0.9,R))                          459
      HNEW = SIGN(AMAX1(HNEW,FOURU*ABS(X)),H)                      460
  465 H = HNEW                                                     461
      RETURN                                                       462
C     ***          END BLOCK 4          ***                        463
      END                                                          464
```

```
      SUBROUTINE INTRP(X,Y,XOUT,YOUT,YPOUT,NEQN,KOLD,PHI,PSI)          1
C                                                                       2
C   THE METHODS IN SUBROUTINE STEP APPROXIMATE THE SOLUTION NEAR X      3
C   BY A POLYNOMIAL.  SUBROUTINE INTRP APPROXIMATES THE SOLUTION AT     4
C   XOUT BY EVALUATING THE POLYNOMIAL THERE.  INFORMATION DEFINING THIS 5
C   POLYNOMIAL IS PASSED FROM STEP SO INTRP CANNOT BE USED ALONE.       6
C                                                                       7
C   THIS CODE IS COMPLETELY EXPLAINED AND DOCUMENTED IN THE TEXT        8
C   COMPUTER SOLUTION OF ORDINARY DIFFERENTIAL EQUATIONS: THE INITIAL   9
C   VALUE PROBLEM BY L. F. SHAMPINE AND M. K. GORDON.                  10
C                                                                      11
C   INPUT TO INTRP --                                                  12
C                                                                      13
C   THE USER PROVIDES STORAGE IN THE CALLING PROGRAM FOR THE ARRAYS IN 14
C   THE CALL LIST                                                      15
C      DIMENSION Y(NEQN),YOUT(NEQN),YPOUT(NEQN),PHI(NEQN,16),PSI(12)   16
C   AND DEFINES                                                        17
C      XOUT -- POINT AT WHICH SOLUTION IS DESIRED.                     18
C   THE REMAINING PARAMETERS ARE DEFINED IN STEP   AND PASSED TO  INTRP 19
C   FROM THAT SUBROUTINE                                               20
C                                                                      21
C   OUTPUT FROM INTRP --                                               22
C                                                                      23
C      YOUT(*) -- SOLUTION AT XOUT                                     24
C      YPOUT(*) -- DERIVATIVE OF SOLUTION AT XOUT                      25
C   THE REMAINING PARAMETERS ARE RETURNED UNALTERED FROM THEIR INPUT   26
C   VALUES.  INTEGRATION WITH STEP MAY BE CONTINUED.                   27
C                                                                      28
      DIMENSION G(13),W(13),RHO(13)                                   29
      DATA G(1)/1.0/,RHO(1)/1.0/                                      30
C                                                                      31
```

```fortran
      HI = XOUT - X
      KI = KOLD + 1
      KIP1 = KI + 1
C
C     INITIALIZE W(*) FOR COMPUTING G(*)
C
      DO 5 I = 1,KI
        TEMP1 = I
    5   W(I) = 1.0/TEMP1
      TERM = 0.0
C
C     COMPUTE G(*)
C
      DO 15 J = 2,KI
        JM1 = J - 1
        PSIJM1 = PSI(JM1)
        GAMMA = (HI + TERM)/PSIJM1
        ETA = HI/PSIJM1
        LIMIT1 = KIP1 - J
        DO 10 I = 1,LIMIT1
   10     W(I) = GAMMA*W(I) - ETA*W(I+1)
        G(J) = W(1)
        RHO(J) = GAMMA*RHO(JM1)
   15   TERM = PSIJM1
C
C     INTERPOLATE
C
      DO 20 L = 1,NEQN
   20   YOUT(L) = 0.0
      DO 30 J = 1,KI
        I = KIP1 - J
        TEMP2 = G(I)
        TEMP3 = RHO(I)
```

```
      DO 25 L = 1, NEQN
         YOUT(L) = YOUT(L) + TEMP2*PHI(L,I)
25       YPOUT(L) = YPOUT(L) + TEMP3*PHI(L,I)
30    CONTINUE
      DO 35 L = 1, NEQN
35       YOUT(L) = Y(L) + HI*YOUT(L)
      RETURN
      END
```

```
      SUBROUTINE MACHIN(U)
C
C  U IS THE SMALLEST POSITIVE NUMBER SUCH THAT (1.0+U) .GT. 1.0 .
C  U IS COMPUTED APPROXIMATELY AS A POWER OF 1./2.
C
C  THIS CODE IS COMPLETELY EXPLAINED AND DOCUMENTED IN THE TEXT
C  COMPUTER SOLUTION OF ORDINARY DIFFERENTIAL EQUATIONS:  THE INITIAL
C  VALUE PROBLEM BY L. F. SHAMPINE AND M. K. GORDON.
C
      HALFU = 0.5
50    TEMP1 = 1.0 + HALFU
      IF(TEMP1 .LE. 1.0) GO TO 100
      HALFU = 0.5*HALFU
      GO TO 50
100   U = 2.0*HALFU
      RETURN
      END
```

USEFUL EXTENSIONS OF THE CODES

The design of the basic codes emphasizes portability, ease of use, and ease of understanding. Because of these goals we omitted several useful features which will be discussed in this section. None of the changes that we suggest affects the performance of the codes, just their flexibility and ease of use. The modified codes use only standard FORTRAN but one is *not* acceptable to the WATFOR compiler and possibly to other compilers. Where the changes are straightforward we just discuss what needs to be altered in the basic codes; where the changes are complex we provide complete listings of the modified versions.

The basic codes have two principal limitations. DE allows no more than 20 equations and is dimensioned to handle up to 20 equations. If we want to work with more equations, a different version of the code must be used. If we want to work with fewer equations, this is being a bit wasteful of storage. STEP does not share this limitation since the user supplies arrays of appropriate dimension for each problem and there is no restriction on the number of equations. Both codes retain variables internally between calls. Versions which do not allow this local retention are needed for computing with overlays and for switching back and forth between problems without restarting. These capabilities are probably unimportant in the classroom but are convenient for production computing. The use of a new driver, ODE, and slightly modified versions of DE and STEP will remove these limitations.

A number of the arrays in DE and STEP have a size which depends on the number of equations. One way to allow variable dimensions is to put these arrays into the call list. This is incompatible with the design of DE, which is intended to be as easy to use as is possible. Another way is for the user to pass in a work array which is then split up into the various arrays that are required. The same possibilities handle the difficulties of overlays and switching equations. Overlays are associated with very large programs, so large that they cannot fit into the rapid access storage of the computer. Programs of this kind are written in such a way that the computer executes them in portions, bringing into the main storage area from peripheral storage a portion that will fit and "overlaying" it on the preceding portion. Just how this is done depends on the operating system. The difficulty for DE and STEP in such an environment is that the variables stored internally between calls may be overwritten by the new portion brought in. There is a similar difficulty when several systems of equations are being alternately solved. Information stored internally when solving one system is overwritten by the code itself when solving a second system. It is impossible to go back to integrating the first system without restarting, but this can get expensive. Since standard FORTRAN specifies that all arguments in the call list will be saved in an overlay, and since they are available for the user to save when switching problems, we can overcome these difficulties at the same time we provide for variable dimensions.

Subroutine STEP already allows a variable number of equations; its only limitation is the local retention of variables between calls. Since relatively few extra parameters are involved it is easier to augment the call list than to introduce work arrays and a driver for dividing the arrays. The modified STEP is accessed by

```
      CALL  STEP(X,Y,F,NEQN,H,EPS,WT,START,
     1    HOLD,K,KOLD,CRASH,PHI,P,YP,PSI,
     2    ALPHA,BETA,SIG,V,W,G,PHASE1,NS,NORND)
```

To distinguish this version from the basic code, we shall refer to this extension as "STEP with the full call list." The only change internal to the code is the deletion of the statement

```
      DATA  G(1),G(2),SIG(1)/1.0,0.5,1.0/
```

and the insertion in Block 0 of the statements

```
   15    CRASH = .FALSE.
         G(1) = 1.0
         G(2) = 0.5
         SIG(1) = 1.0
```

With the exception of providing storage in the calling program for

$$ALPHA(12), BETA(12), SIG(13), V(12), W(12), G(13)$$

and declaring PHASE1 and NORND as logical variables, STEP with the full call list is used in exactly the same way as the basic STEP.

Augmenting the call list to this degree is unacceptable in DE. Instead, we have added two work arrays to the call list and another level of sub-routine to allocate storage in these arrays. (This use of virtual storage is not acceptable to the WATFOR compiler and the code ODE will not run under it.) Thus we propose a new driver

ODE(F,NEQN,Y,T,TOUT,RELERR,ABSERR,IFLAG,WORK,IWORK)

which, as far as the user is concerned, replaces DE. The only distinction the user sees is that he must supply $WORK(100 + 21 \cdot NEQN)$ and $IWORK(5)$ in the calling program and cannot alter their output values. Otherwise ODE is used exactly like DE. ODE divides the work arrays into the separate pieces needed by the other subroutines and calls a modified version of DE which directs the integration. The first one hundred words of WORK and all of IWORK hold, respectively, the real and integer variables internal to DE and STEP. The remaining $21 \cdot NEQN$ words of WORK hold the arrays

YY(NEQN),WT(NEQN),P(NEQN),YP(NEQN),YPOUT(NEQN),
PHI(NEQN,16),

in that order. The changes in DE are a matter of increasing the call lists and adjusting the DIMENSION statements. No additional changes are necessary for STEP with the full call list. ODE and the modified DE are listed below.

To integrate several systems of equations with either of the modified codes, the user need only supply separate arguments for each system. For example, to alternately integrate two systems with ODE the user might call the subroutines as follows:

CALL ODE(F1,NEQN1,Y1,T1,TOUT1,RE1,
1   AE1,IFLAG1,WORK1,IWORK1)

and

CALL ODE(F2,NEQN2,Y2,T2,TOUT2,RE2,
1   AE2,IFLAG2,WORK2,IWORK2).

In this way, information pertinent to one system is preserved while integrating the other.

```
      SUBROUTINE ODE(F,NEQN,Y,T,TOUT,RELERR,ABSERR,IFLAG,WORK,IWORK)    1
                                                                        2
C     SUBROUTINE ODE  INTEGRATES A SYSTEM OF  NEQN  FIRST ORDER         3
C     ORDINARY DIFFERENTIAL EQUATIONS OF THE FORM                       4
C                DY(I)/DT = F(T,Y(1),Y(2),...,Y(NEQN))                  5
C                Y(I) GIVEN AT  T .                                     6
C                                                                       7
C     THE SUBROUTINE INTEGRATES FROM  T  TO  TOUT .  ON RETURN THE      8
C     PARAMETERS IN THE CALL LIST ARE SET FOR CONTINUING THE INTEGRATION. 9
C     THE USER HAS ONLY TO DEFINE A NEW VALUE  TOUT  AND CALL  ODE  AGAIN. 10
C                                                                       11
C     THE DIFFERENTIAL EQUATIONS ARE ACTUALLY SOLVED BY A SUITE OF CODES 12
C     DE ,  STEP , AND  INTRP .  ODE  ALLOCATES VIRTUAL STORAGE IN THE  13
C     ARRAYS  WORK  AND  IWORK  AND CALLS  DE .  DE  IS A SUPERVISOR WHICH 14
C     DIRECTS THE SOLUTION.  IT CALLS ON THE ROUTINES  STEP  AND  INTRP 15
C     TO ADVANCE THE INTEGRATION AND TO INTERPOLATE AT OUTPUT POINTS.   16
C     STEP  USES A MODIFIED DIVIDED DIFFERENCE FORM OF THE ADAMS PECE   17
C     FORMULAS AND LOCAL EXTRAPOLATION.  IT ADJUSTS THE ORDER AND STEP  18
C     SIZE TO CONTROL THE LOCAL ERROR PER UNIT STEP IN A GENERALIZED    19
C     SENSE.  NORMALLY EACH CALL TO  STEP  ADVANCES THE SOLUTION ONE STEP 20
C     IN THE DIRECTION OF  TOUT .  FOR REASONS OF EFFICIENCY  DE         21
C     INTEGRATES BEYOND  TOUT  INTERNALLY, THOUGH NEVER BEYOND          22
C     T+10*(TOUT-T) , AND CALLS  INTRP  TO INTERPOLATE THE SOLUTION AT  23
C     TOUT .  AN OPTION IS PROVIDED TO STOP THE INTEGRATION AT  TOUT  BUT 24
C     IT SHOULD BE USED ONLY IF IT IS IMPOSSIBLE TO CONTINUE THE        25
C     INTEGRATION BEYOND  TOUT .                                        26
C                                                                       27
C     THIS CODE IS COMPLETELY EXPLAINED AND DOCUMENTED IN THE TEXT      28
C     COMPUTER SOLUTION OF ORDINARY DIFFERENTIAL EQUATIONS:  THE INITIAL 29
C     VALUE PROBLEM  BY L. F. SHAMPINE AND M. K. GORDON.               30
```

```
C     THE PARAMETERS REPRESENT:                                              31
C     F -- SUBROUTINE F(T,Y,YP) TO EVALUATE DERIVATIVES YP(I)=DY(I)/DT       32
C     NEQN -- NUMBER OF EQUATIONS TO BE INTEGRATED                           33
C     Y(*) -- SOLUTION VECTOR AT T                                           34
C     T -- INDEPENDENT VARIABLE                                              35
C     TOUT -- POINT AT WHICH SOLUTION IS DESIRED                             36
C     RELERR,ABSERR -- RELATIVE AND ABSOLUTE ERROR TOLERANCES FOR LOCAL      37
C        ERROR TEST.  AT EACH STEP THE CODE REQUIRES                         38
C        ABS(LOCAL ERROR) .LE. ABS(Y)*RELERR + ABSERR                        39
C     FOR EACH COMPONENT OF THE LOCAL ERROR AND SOLUTION VECTORS             40
C     IFLAG -- INDICATES STATUS OF INTEGRATION                               41
C     WORK(*),IWORK(*) -- ARRAYS TO HOLD INFORMATION INTERNAL TO CODE        42
C        WHICH IS NECESSARY FOR SUBSEQUENT CALLS                             43
C                                                                            44
C     FIRST CALL TO ODE --                                                   45
C                                                                            46
C     THE USER MUST PROVIDE STORAGE IN HIS CALLING PROGRAM FOR THE ARRAYS    47
C     IN THE CALL LIST                                                       48
C        Y(NEQN), WORK(100+21*NEQN) IWORK(5),                                49
C     DECLARE F IN AN EXTERNAL STATEMENT, SUPPLY THE SUBROUTINE              50
C     F(T,Y,YP) TO EVALUATE                                                  51
C        DY(I)/DT = YP(I) = F(T,Y(1),Y(2),...,Y(NEQN))                       52
C     AND INITIALIZE THE PARAMETERS:                                         53
C     NEQN -- NUMBER OF EQUATIONS TO BE INTEGRATED                           54
C     Y(*) -- VECTOR OF INITIAL CONDITIONS                                   55
C     T -- STARTING POINT OF INTEGRATION                                     56
C     TOUT -- POINT AT WHICH SOLUTION IS DESIRED                             57
C     RELERR,ABSERR -- RELATIVE AND ABSOLUTE LOCAL ERROR TOLERANCES          58
C     IFLAG -- +1, -1.  INDICATOR TO INITIALIZE THE CODE.  NORMAL INPUT      59
C        IS +1.  THE USER SHOULD SET IFLAG=-1 ONLY IF IT IS                  60
C        IMPOSSIBLE TO CONTINUE THE INTEGRATION BEYOND TOUT                  61
C     ALL PARAMETERS EXCEPT F, NEQN AND TOUT MAY BE ALTERED BY THE           62
C     CODE ON OUTPUT SO MUST BE VARIABLES IN THE CALLING PROGRAM.            63
C                                                                            64
C     OUTPUT FROM ODE --                                                     65
C                                                                            66
```

```
C      NEQN -- UNCHANGED                                                    67
C      Y(*) -- SOLUTION AT T                                               68
C      T -- LAST POINT REACHED IN INTEGRATION.  NORMAL RETURN HAS          69
C           T = TOUT.                                                      70
C      TOUT -- UNCHANGED                                                   71
C      RELERR,ABSERR -- NORMAL RETURN HAS TOLERANCES UNCHANGED.  IFLAG=3   72
C           SIGNALS TOLERANCES INCREASED                                   73
C      IFLAG = 2 -- NORMAL RETURN.  INTEGRATION REACHED TOUT               74
C            = 3 -- INTEGRATION DID NOT REACH TOUT BECAUSE ERROR           75
C                   TOLERANCES TOO SMALL.  RELERR , ABSERR INCREASED       76
C                   APPROPRIATELY FOR CONTINUING                           77
C            = 4 -- INTEGRATION DID NOT REACH TOUT  BECAUSE MORE THAN      78
C                   500 STEPS NEEDED                                       79
C            = 5 -- INTEGRATION DID NOT REACH TOUT BECAUSE EQUATIONS       80
C                   APPEAR TO BE STIFF                                     81
C            = 6 -- INVALID INPUT PARAMETERS (FATAL ERROR)                 82
C           THE VALUE OF IFLAG IS RETURNED NEGATIVE WHEN THE INPUT         83
C           VALUE IS NEGATIVE AND THE INTEGRATION DOES NOT REACH TOUT ,    84
C           I.E.,-3,-4,-5.                                                 85
C      WORK(*),IWORK(*) -- INFORMATION GENERALLY OF NO INTEREST TO THE     86
C           USER BUT NECESSARY FOR SUBSEQUENT CALLS.                       87
C                                                                          88
C      SUBSEQUENT CALLS TO ODE --                                         89
C                                                                          90
C      SUBROUTINE ODE RETURNS WITH ALL INFORMATION NEEDED TO CONTINUE      91
C      THE INTEGRATION.  IF THE INTEGRATION REACHED TOUT THE USER NEED     92
C      ONLY DEFINE A NEW TOUT AND CALL AGAIN.  IF THE INTEGRATION DID NOT  93
C      REACH TOUT AND THE USER WANTS TO CONTINUE HE JUST CALLS AGAIN.      94
C      THE OUTPUT VALUE OF IFLAG IS THE APPROPRIATE INPUT VALUE FOR        95
C      SUBSEQUENT CALLS.  THE ONLY SITUATION IN WHICH IT SHOULD BE ALTERED 96
C      IS TO STOP THE INTEGRATION INTERNALLY AT THE NEW TOUT I.E.,         97
C      CHANGE OUTPUT IFLAG=2 TO INPUT IFLAG=-2 . ERROR TOLERANCES MAY      98
C      BE CHANGED BY THE USER BEFORE CONTINUING.  ALL OTHER PARAMETERS MUST 99
C      REMAIN UNCHANGED.                                                   100
C                                                                          101
```

```
C*******************************************************
C*  SUBROUTINES  DE  AND  STEP  CONTAIN MACHINE DEPENDENT CONSTANTS.  *
C*  BE SURE THEY ARE SET BEFORE USING  ODE .                          *
C*******************************************************
C
      LOGICAL START PHASE1 NORND
      DIMENSION Y(NEQN),WORK(1),IWORK(5)
      EXTERNAL F
      DATA IALPHA,IBETA,ISIG,IV,IG,IPHASE,IPSI,IX,IH,IHOLD,ISTART,
     1  ITOLD,IDELSN/1,13,25,38,50,62,75,76,88,89,90,91,92,93/
      IYY = 100
      IWT = IYY + NEQN
      IP = IWT + NEQN
      IYP = IP + NEQN
      IYPOUT = IYP + NEQN
      IPHI = IYPOUT + NEQN
      IF(IABS(IFLAG) .EQ. 1) GO TO 1
      START = WORK(ISTART) .GT. 0.0
      PHASE1 = WORK(IPHASE) .GT. 0.0
      NORND = IWORK(2) .NE. -1
    1 CALL DE(F NEQN Y T TOUT RELERR ABSERR IFLAG WORK(IYY)
     1  WORK(IWT) WORK(IP) WORK(IYP) WORK(IYPOUT) WORK(IPHI)
     2  WORK(IALPHA) WORK(IBETA) WORK(ISIG) WORK(IV) WORK(IW) WORK(IG)
     3  PHASE1 WORK(IPSI) WORK(IX) WORK(IH) WORK(IHOLD) START
     4  WORK(ITOLD) WORK(IDELSN) IWORK(1) NORND IWORK(3) IWORK(4)
     5  IWORK(5))
      WORK(ISTART) = -1.0
      IF(START) WORK(ISTART) = 1.0
      WORK(IPHASE) = -1.0
      IF(PHASE1) WORK(IPHASE) = 1.0
      IWORK(2) = -1
      IF(NORND) IWORK(2) = 1
      RETURN
      END
```

```
      SUBROUTINE DE(F,NEQN,Y,T,TOUT,RELERR,ABSERR,IFLAG,                 1
     1 YY,WT,P,YP,YPOUT,PHI,ALPHA,BETA,SIG,V,W,G,PHASE1,PSI,X,H,HOLD,    2
     2 START,TOLD,DELSGN,NS,NORND,K,KOLD,ISNOLD)                         3
C                                                                        4
C   ODE MERELY ALLOCATES STORAGE FOR  DE  TO RELIEVE THE USER OF THE     5
C   INCONVENIENCE OF A LONG CALL LIST.  CONSEQUENTLY  DE  IS USED AS     6
C   DESCRIBED IN THE COMMENTS FOR  ODE .                                 7
C                                                                        8
C   THIS CODE IS COMPLETELY EXPLAINED AND DOCUMENTED IN THE TEXT         9
C   COMPUTER SOLUTION OF ORDINARY DIFFERENTIAL EQUATIONS:  THE INITIAL   10
C   VALUE PROBLEM  BY L. F. SHAMPINE AND M. K. GORDON.                   11
C                                                                        12
      LOGICAL STIFF,CRASH,START,PHASE1,NORND                             13
      DIMENSION Y(NEQN),YY(NEQN),WT(NEQN),PHI(NEQN,16),P(NEQN),YP(NEQN), 14
     1 YPOUT(NEQN),PSI(12),ALPHA(12),BETA(12),SIG(13),V(12),W(12),G(13)  15
      EXTERNAL F                                                         16
C                                                                        17
C***********************************************************************  18
C*   THE ONLY MACHINE DEPENDENT CONSTANT IS BASED ON THE MACHINE UNIT  *  19
C*   ROUNDOFF ERROR  U  WHICH IS THE SMALLEST POSITIVE NUMBER SUCH THAT*  20
C*   1.0+U .GT. 1.0 .  U  MUST BE CALCULATED AND  FOURU=4.0*U  INSERTED*  21
C*   IN THE FOLLOWING DATA STATEMENT BEFORE USING  DE .  THE ROUTINE   *  22
C*   MACHIN CALCULATES  U .  FOURU  AND  TWOU=2.0*U  MUST ALSO BE      *  23
C*   INSERTED IN SUBROUTINE  STEP  BEFORE CALLING  DE .               *  24
C        DATA FOURU /           /                                        25
C***********************************************************************  26
```

```
C     THE CONSTANT MAXNUM IS THE MAXIMUM NUMBER OF STEPS ALLOWED IN ONE
C     CALL TO DE.  THE USER MAY CHANGE THIS LIMIT BY ALTERING THE
C     FOLLOWING STATEMENT
      DATA MAXNUM/500/
C
C                       ***                ***
C     TEST FOR IMPROPER PARAMETERS
C
      IF(NEQN .LT. 1) GO TO 10
      IF(T .EQ. TOUT) GO TO 10
      IF(RELERR .LT. 0.0  .OR.  ABSERR .LT. 0.0) GO TO 10
      EPS = AMAX1(RELERR,ABSERR)
      IF(EPS .LE. 0.0) GO TO 10
      IF(IFLAG .EQ. 0) GO TO 10
      ISN = ISIGN(1,IFLAG)
      IFLAG = IABS(IFLAG)
      IF(IFLAG .EQ. 1) GO TO 20
      IF(T .NE. TOLD) GO TO 10
      IF(IFLAG .GE. 2  .AND.  IFLAG .LE. 5 ) GO TO 20
   10 IFLAG = 6
      RETURN
C
C     ON EACH CALL SET INTERVAL OF INTEGRATION AND COUNTER FOR NUMBER OF
C     STEPS.  ADJUST INPUT ERROR TOLERANCES TO DEFINE WEIGHT VECTOR FOR
C     SUBROUTINE STEP
C
   20 DEL = TOUT - T
      ABSDEL = ABS(DEL)
      TEND = T + 10.0*DEL
      IF(ISN .LT. 0) TEND = TOUT
      NOSTEP = 0
      KLE4 = 0
```

```
   60       STIFF = .FALSE.
   61       RELEPS = RELERR/EPS
   62       ABSEPS = ABSERR/EPS
   63       IF(IFLAG .EQ. 1) GO TO 30
   64       IF(ISNOLD .LT. 0) GO TO 30
   65       IF(DELSGN*DEL .GT. 0.0) GO TO 50
   66 C
   67 C   ON START AND RESTART ALSO SET WORK VARIABLES X AND YY(*), STORE THE
   68 C   DIRECTION OF INTEGRATION AND INITIALIZE THE STEP SIZE
   69 C
   70    30 START = .TRUE.
   71       X = T
   72       DO 40 L = 1,NEQN
   73    40 YY(L) = Y(L)
   74       DELSGN = SIGN(1.0,DEL)
   75       H = SIGN(AMAX1(ABS(TOUT-X),FOURU*ABS(X)),TOUT-X)
   76 C
   77 C   IF ALREADY PAST OUTPUT POINT, INTERPOLATE AND RETURN
   78 C
   79    50 IF(ABS(X-T) .LT. ABSDEL) GO TO 60
   80       CALL INTRP(X,YY,TOUT,Y,YPOUT,NEQN,KOLD,PHI,PSI)
   81       IFLAG = 2
   82       T = TOUT
   83       TOLD = T
   84       ISNOLD = ISN
   85       RETURN
   86 C
   87 C   IF CANNOT GO PAST OUTPUT POINT AND SUFFICIENTLY CLOSE,
   88 C   EXTRAPOLATE AND RETURN
   89 C
   90    60 IF(ISN .GT. 0  .OR.  ABS(TOUT-X) .GE. FOURU*ABS(X)) GO TO 80
   91       H = TOUT - X
   92       CALL F(X,YY,YP)
   93       DO 70 L = 1,NEQN
```

```
70    Y(L) = YY(L) + H*YP(L)
      IFLAG = 2
      T = TOUT
      TOLD = T
      ISNOLD = ISN
      RETURN
C
C     TEST FOR TOO MANY STEPS
C
80    IF(NOSTEP .LT. MAXNUM) GO TO 100
      IFLAG = ISN*4
      IF(STIFF) IFLAG = ISN*5
      DO 90 L = 1,NEQN
90    Y(L) = YY(L)
      T = X
      TOLD = T
      ISNOLD = 1
      RETURN
C
C     LIMIT STEP SIZE, SET WEIGHT VECTOR AND TAKE A STEP
C
100   H = SIGN(AMIN1(ABS(H),ABS(TEND-X)),H)
      DO 110 L = 1,NEQN
110   WT(L) = RELEPS*ABS(YY(L)) + ABSEPS
      CALL STEP(F,NEQN,YY,X,H,EPS,WT,START,
     1   HOLD,K,KOLD,CRASH,PHI,P,YP,PSI,
     2   ALPHA,BETA,SIG,V,W,G,PHASE1,NS,NORND)
C
C     TEST FOR TOLERANCES TOO SMALL
C
      IF(.NOT.CRASH) GO TO 130
      IFLAG = ISN*3
      RELERR = EPS*RELEPS
```

```fortran
      ABSERR = EPS*ABSEPS
      DO 120 L = 1,NEQN
  120 Y(L) = YY(L)
      T = X
      TOLD = T
      ISNOLD = 1
      RETURN
C
C  AUGMENT COUNTER ON NUMBER OF STEPS AND TEST FOR STIFFNESS
C
  130 NOSTEP = NOSTEP + 1
      KLE4 = KLE4 + 1
      IF(KOLD .GT. 4) KLE4 = 0
      IF(KLE4 .GE. 50) STIFF = .TRUE.
      GO TO 50
      END
```

127
128
129
130
131
132
133
134
135
136
137
138
139
140
141
142

In Chapter 7 we discussed the estimation of the residual of a computed solution and its significance as a measure of the quality of a solution. The codes furnish a solution, $y_I(x)$, which satisfies

$$y_I'(x) = f(x, y_I(x)) + r(x)$$
$$= (1 + \rho(x))f(x, y_I(x)) + \tau(x).$$

We shall estimate the smallest number μ such that the relative error ρ and absolute error τ are bounded by μ throughout the integration:

$$\mu = \max_{[a,b]} \frac{|r(x)|}{1 + |f(x, y_I(x))|}.$$

For systems of equations we use the largest value found in this way for any of the equations. Since the residual vanishes at each mesh point, an adequate estimate of μ can be obtained by evaluating the ratio

$$\frac{|r(x)|}{1 + |f(x, y_I(x))|}$$

at the midpoint of each step. To modify DE to do this and return the estimate RES, only the changes indicated in the following are required:

```
       SUBROUTINE DE(F,NEQN,Y,T,TOUT,RELERR,ABSERR,IFLAG,RES)
           .
           .
           .
       ABSEPS = ABSERR/EPS
       IF(IFLAG .GT. 1) GO TO 25
       RES = 0.0
       GO TO 30
25     IF(ISNOLD .LT. 0) GO TO 30
           .
           .
           .
       IF(KLE4 .GT. 50) STIFF = .TRUE.
       XMID = X - 0.5*HOLD
       CALL INTRP(X,YY,XMID,Y,YPOUT,NEQN,KOLD,PHI,PSI)
       CALL F(XMID,Y,YP)
       DO 150 L = 1,NEQN
150      RES = AMAX1(RES,ABS(YPOUT(L)-YP(L))/(1.0+ABS(YP(L))))
       GO TO 50
       END
```

This estimate seems to be quite satisfactory but, as it also corresponds to one extra derivative evaluation per step, it increases the cost by about 50%. Since these codes apparently always yield solutions with residuals of the size requested, it is difficult to justify going to this expense routinely.

Some users will want to detect an attempt to compute a vanishing solution with a pure relative error test. The only situation in which this is at all likely is if an initial condition is exactly zero. However, it can occur if the computed solution is exactly zero at a mesh point. Normally, this will be due to an underflow. To be consistent in the use of IFLAG we should use IFLAG = ± 6 to indicate an error of this kind and use IFLAG = 7 for invalid input. There is no simple, automatic remedy for such errors since an appropriate absolute error tolerance depends on the scale of the dependent variables. To fully protect the code, insert just before statement 10

```
IF(IFLAG .GE. 2 .AND. IFLAG .LE. 6) GO TO 20
```

replace statement 10 with

```
10      IFLAG = 7
```

replace statement 110 with

```
        WT(L) = RELEPS * ABS(YY(L)) + ABSEPS
110     IF(WT(L) .LE. 0.0) GO TO 140
```

and insert before the END statement

```
140     IFLAG = ISN*6
        DO 150 L = 1, NEQN
150        Y(L) = YY(L)
        T = X
        TOLD = T
        ISNOLD = 1
        RETURN
```

A ROOT-SOLVER

Quite commonly a root-solving code is needed in conjunction with a differential equation solver. Several examples are discussed in Chapter 12. In reference [25] an excellent code, ZEROIN, is fully explained and documented. It computes a solution of $F(t) = 0$ which lies in a specified interval $[B, C]$. This code requires a FUNCTION subprogram for evaluating $F(t)$ anywhere in this interval. For the purposes associated with the solution of differential equations this requirement is awkward because the value of the function is usually obtained by interpolation using INTRP or by the solution of a differential equation using DE or STEP. While it is possible to program the evaluation of the function as a FUNCTION subprogram using COMMON, it is clumsy at best and the inexperienced may find it difficult. We have modified ZEROIN so that it returns to the calling program every time a function value is needed. The user computes it and calls the modified code, ROOT, again. The modifications are quite minor so we shall describe how to use ROOT, give a listing for it, and refer the reader to [25] for its general documentation. It will be seen in Chapter 12 that the design change is very convenient in this context.

Subroutine ROOT(T,FT,B,C,RELERR,ABSERR,IFLAG) calculates a root of the nonlinear equation $F(T) = 0$. The user supplies an interval $[B,C]$ such that $F(B)$ and $F(C)$ are of opposite signs. Assuming that F is continuous on this interval, there is a root of the equation bracketed by B and C. Successive approximations to the root are computed and the interval $[B,C]$ is shrunk while still bracketing a root. A successful computation results in

$$\left|\frac{B - C}{2}\right| \leq |B|*RELERR + ABSERR$$

and $F(B)$, $F(C)$ of opposite signs. B is the better approximate root in the sense that $|F(B)| \leq |F(C)|$. The quantities RELERR and ABSERR are, respectively, the relative and absolute error tolerances supplied by the user.

The value $F(T)$ is required for each approximation T. To get it, the code returns control to the calling program with IFLAG negative and a value T at which F is to be evaluated. The user computes $FT = F(T)$ and calls ROOT again. (Because of the way this is done, ROOT retains variables internally between calls.) ROOT continues shrinking the interval until it converges, or detects some abnormal condition. In either circumstance it returns with a positive value of IFLAG to report what happened.

IFLAG must be set to a positive value before calling the code the first time in order to initialize it. Thereafter, if the code returns with a negative value for IFLAG, the user is to evaluate $FT = F(T)$ and call it again. If it returns with a positive value, the code is finished. The parameters

T,FT,B,C, and **IFLAG** must be variables in the calling program since they are used for both input and output. As a simple example we solve $\exp(-t) - t = F(t) = 0$. Since $F(0) > 0$ and $F(1) < 0$ there is a root in $[0, 1]$. The following program will compute this root.

```
        B = 0.0
        C = 1.0
        RELERR = 1.0E-6
        ABSERR = 1.0E-6
        IFLAG = 1
  1     CALL  ROOT(T,FT,B,C,RELERR,ABSERR,IFLAG)
        IF(IFLAG .GT. 0) GO TO 2
        FT = EXP(-T) - T
        GO TO 1
  2     CONTINUE
```

The final interval (B, C) is $(0.5671433, 0.5671417)$ and IFLAG is 1.

   This brief discussion is intended to show how to use the modified code in the most commonly occurring situations. The comments, beginning on the next page, which form the prologue to subroutine ROOT will assist in using it. The reader will benefit from the full explanations for ZEROIN in reference [25] if he is going to use the code often.

```
      SUBROUTINE ROOT(T,FT,B,C,RELERR,ABSERR,IFLAG)               1
C                                                                 2
C     ROOT COMPUTES A ROOT OF THE NONLINEAR EQUATION F(X)=0       3
C     WHERE F(X) IS A CONTINUOUS REAL FUNCTION OF A SINGLE REAL   4
C     VARIABLE X.  THE METHOD USED IS A COMBINATION OF BISECTION  5
C     AND THE SECANT RULE.                                        6
C                                                                 7
C     NORMAL INPUT CONSISTS OF A CONTINUOUS FUNCTION F AND AN     8
C     INTERVAL (B,C) SUCH THAT F(B)*F(C).LE.0.0.  EACH ITERATION  9
C     FINDS NEW VALUES OF B AND C SUCH THAT THE INTERVAL (B,C) IS 10
C     SHRUNK AND F(B)*F(C).LE.0.0.  THE STOPPING CRITERION IS     11
C                                                                 12
C         ABS(B-C).LE.2.0*(RELERR*ABS(B)+ABSERR)                  13
C                                                                 14
C     WHERE RELERR=RELATIVE ERROR AND ABSERR=ABSOLUTE ERROR ARE   15
C     INPUT QUANTITIES.  SET THE FLAG, IFLAG, POSITIVE TO INITIALIZE 16
C     THE COMPUTATION.  AS B,C AND IFLAG ARE USED FOR BOTH INPUT AND 17
C     OUTPUT, THEY MUST BE VARIABLES IN THE CALLING PROGRAM.      18
C                                                                 19
C     IF 0 IS A POSSIBLE ROOT, ONE SHOULD NOT CHOOSE ABSERR=0.0.  20
C                                                                 21
C     THE OUTPUT VALUE OF B IS THE BETTER APPROXIMATION TO A ROOT 22
C     AS B AND C ARE ALWAYS REDEFINED SO THAT ABS(F(B)).LE.ABS(F(C)). 23
C                                                                 24
```

```
C     TO SOLVE THE EQUATION.  ROOT MUST EVALUATE F(X) REPEATEDLY.  THIS     25
C     IS DONE IN THE CALLING PROGRAM.  WHEN AN EVALUATION OF F IS           26
C     NEEDED AT T ROOT RETURNS WITH IFLAG NEGATIVE.  EVALUATE FT=F(T)       27
C     AND CALL ROOT AGAIN.  DO NOT ALTER IFLAG.                             28
C                                                                          29
C     WHEN THE COMPUTATION IS COMPLETE, ROOT RETURNS TO THE CALLING         30
C     PROGRAM WITH IFLAG POSITIVE:                                          31
C                                                                          32
C        IFLAG=1  IF F(B)*F(C).LT.0 AND THE STOPPING CRITERION IS MET.      33
C                                                                          34
C             =2  IF A VALUE B IS FOUND SUCH THAT THE COMPUTED VALUE        35
C                 F(B) IS EXACTLY ZERO.  THE INTERVAL (B,C) MAY NOT         36
C                 SATISFY THE STOPPING CRITERION.                           37
C                                                                          38
C             =3  IF ABS(F(B)) EXCEEDS THE INPUT VALUES ABS(F(B)),          39
C                 ABS(F(C)).  IN THIS CASE IT IS LIKELY THAT B IS CLOSE     40
C                 TO A POLE OF F.                                           41
C                                                                          42
C             =4  IF NO ODD ORDER ROOT WAS FOUND IN THE INTERVAL.  A        43
C                 LOCAL MINIMUM MAY HAVE BEEN OBTAINED.                     44
C                                                                          45
C             =5  IF TOO MANY FUNCTION EVALUATIONS WERE MADE.               46
C                 (AS PROGRAMMED, 500 ARE ALLOWED.)                         47
C                                                                          48
C     THIS CODE IS A MODIFICATION OF THE CODE ZEROIN  WHICH IS COMPLETELY   49
C     EXPLAINED AND DOCUMENTED IN THE TEXT NUMERICAL COMPUTING: AN          50
C     INTRODUCTION BY L. F. SHAMPINE AND R. C. ALLEN.                       51
C                                                                          52
```

```
C***********************************************************
C*  THE ONLY MACHINE DEPENDENT CONSTANT IS BASED ON THE MACHINE UNIT  *       53
C*  ROUNDOFF ERROR  U  WHICH IS THE SMALLEST POSITIVE NUMBER SUCH THAT  *     54
C*  1.0+U .GT. 1.0 .  U  MUST BE CALCULATED AND INSERTED IN THE  *            55
C*  FOLLOWING DATA STATEMENT BEFORE USING ROOT .  THE ROUTINE  MACHIN  *      56
C*  CALCULATES  U .                                                  *        57
C*      DATA U /        /                                                     58
C***********************************************************                  59
C                                                                            60
      IF(IFLAG.GE.0) GO TO 100                                               61
      IFLAG=IABS(IFLAG)                                                      62
      GO TO (200,300,400), IFLAG                                            63
  100 RE=AMAX1(RELERR,U)                                                     64
      AE=AMAX1(ABSERR,0.0)                                                   65
      IC=0                                                                   66
      ACBS=ABS(B-C)                                                          67
      A=C                                                                    68
      T=A                                                                    69
      IFLAG=-1                                                               70
      RETURN                                                                 71
  200 FA=FT                                                                  72
      T=B                                                                    73
      IFLAG=-2                                                               74
      RETURN                                                                 75
  300 FB=FT                                                                  76
      FC=FA                                                                  77
      KOUNT=2                                                                78
      FX=AMAX1(ABS(FB),ABS(FC))                                             79
    1 IF(ABS(FC).GE.ABS(FB))GO TO 2                                         80
C                                                                            81
C     INTERCHANGE B AND C SO THAT ABS(F(B)).LE.ABS(F(C)).                   82
C                                                                            83
      A=B                                                                    84
```

```
86          FA=FB
87          B=C
88          FB=FC
89          C=A
90          FC=FA
91        2 CMB=0.5*(C-B)
92          ACMB=ABS(CMB)
93          TOL=RE*ABS(B)+AE
94    C
95    C TEST STOPPING CRITERION AND FUNCTION COUNT.
96    C
97          IF(ACMB.LE.TOL)GO TO 8
98          IF(KOUNT.GE.500)GO TO 12
99    C
100   C CALCULATE NEW ITERATE IMPLICITLY AS B+P/Q
101   C WHERE WE ARRANGE P.GE.O.  THE IMPLICIT
102   C FORM IS USED TO PREVENT OVERFLOW.
103   C
104         P=(B-A)*FB
105         Q=FA-FB
106         IF(P.GE.0.0)GO TO 3
107         P=-P
108         Q=-Q
109   C
110   C UPDATE A, CHECK IF REDUCTION IN THE SIZE OF BRACKETING
111   C INTERVAL IS SATISFACTORY.  IF NOT, BISECT UNTIL IT IS.
112   C
113       3 A=B
114         FA=FB
115         IC=IC+1
116         IF(IC.LT.4)GO TO 4
117         IF(8.0*ACMB.GE.ACBS)GO TO 6
118         IC=0
119         ACBS=ACMB
```

```
C     TEST FOR TOO SMALL A CHANGE.
C
    4 IF(P.GT.ABS(Q)*TOL)GO TO 5
C
C     INCREMENT BY TOLERANCE.
C
      B=B+SIGN(TOL,CMB)
      GO TO 7
C
C     ROOT OUGHT TO BE BETWEEN B AND (C+B)/2.
C
    5 IF(P.GE.CMB*Q)GO TO 6
C
C     USE SECANT RULE.
C
      B=B+P/Q
      GO TO 7
C
C     USE BISECTION.
C
    6 B=0.5*(C+B)
C
C     HAVE COMPLETED COMPUTATION FOR NEW ITERATE B.
C
    7 T=B
      IFLAG=-3
      RETURN
  400 FB=FT
      IF(FB.EQ.0.0)GO TO 9
      KOUNT=KOUNT+1
      IF(SIGN(1.0,FB).NE.SIGN(1.0,FC))GO TO 1
      C=A
```

120
121
122
123
124
125
126
127
128
129
130
131
132
133
134
135
136
137
138
139
140
141
142
143
144
145
146
147
148
149
150
151
152

```
      FC=FA                                                    153
      GO TO 1                                                  154
                                                               155
C                                                              156
C     FINISHED.  SET IFLAG.                                    157
C                                                              158
    8 IF(SIGN(1.0,FB).EQ.SIGN(1.0,FC))GO TO 11                 159
      IF(ABS(FB).GT.FX)GO TO 10                                160
      IFLAG=1                                                  161
      RETURN                                                   162
    9 IFLAG=2                                                  163
      RETURN                                                   164
   10 IFLAG=3                                                  165
      RETURN                                                   166
   11 IFLAG=4                                                  167
      RETURN                                                   168
   12 IFLAG=5                                                  169
      RETURN                                                   170
      END
```

11

CODE PERFORMANCE
AND EVALUATION

The two principal goals of this book are to show how to solve differential equations and to present high quality mathematical software for doing it. The evaluation of software is partly subjective, so in this chapter we raise points relevant to evaluation and comment about what we have done or present computed data. No attempt is made here to compare the codes provided to other codes in the literature, though we have made such comparisons. It is fair to say that our codes are among the very best, but subjective considerations and the computing environment are too important to allow us to say much more. And so we attempt to give the user some idea of what to expect of our codes and to facilitate those comparisons and judgements which he may choose to make.

(i) Documentation of the algorithm: To simply say that a code is a variable step, variable order Adams code is so vague as to be almost useless. Details of the implementation, especially in matters such as starting and selecting the step size and order, cause very significant differences in performance. The algorithms that are employed are completely described in this text, including the heuristics. Because of the teaching aspect of the text we have stressed simplicity and theoretical support in selecting among various algorithmic possibilities.

(ii) Documentation of the code: The codes are intended to be readable and lucid. Where they are not straightforward implementations of the algorithm the reasons are explained. Structure, disciplined FORTRAN, comments, and flow charts are all employed to clarify the coding. The small deviations from standard FORTRAN are common to differential equation solvers and clearly pointed out. The use of machine dependent constants is clearly described.

(iii) Convenience: An extremely important and difficult aspect of code design is to make codes as easy to use as possible but flexible enough to handle all anticipated requirements. This is very subjective, but user response to our code design has been favorable. The driver DE demands little of the user and solves most common problems. Problems with special features must be solved with STEP and INTRP, which are by no means inconvenient. There are a number of important problems which require a code to compute roots of nonlinear equations in conjunction with a differential equation solver. We provide a code that is designed for this purpose. In the next chapter there are many examples of its use.

(iv) Reliability: The codes have been carefully studied and tested on a variety of problems, some of which are reported here. They have been run routinely on an IBM 360/67, PDP-10, and CDC 6600 with various compilers and operating systems and, to a lesser degree, have been run on other machines. Examples have been constructed to exercise the major flow paths in the codes. Some of these are discussed in examples 4, 14, 16, and 18 in this chapter. The codes have been used in both academic and industrial environments. No problems have arisen with the versions presented in Chapter 10. Still, no experienced computor would be so bold as to claim that programs of this complexity are utterly reliable. The best that can reasonably be done is to gain a high degree of confidence and assure the users that: *the authors will respond to constructive criticism of the codes and to documented reports of unsatisfactory performance by the codes on problems for which they are intended.*

(v) Performance: The theory of differential equation solvers is far from complete, so that the understanding of a code's performance requires a study of experimental results. It is only very recently that attempts have been made to collect test problems for this purpose. Two outstanding collections are those of Hull, *et al.* [13] and Krogh [19]. Both have been used extensively in the development and testing of the codes given in this text. In this chapter we present results for Krogh's set to illustrate code performance. They are supplemented by additional computations to show the roles played by various devices in the codes and to illustrate what happens in the case of problems for which the codes are not intended.

Advances in the theory, and the experience of users, may prompt alterations to these codes. Readers are invited to write to the authors for the changes that they endorse.

Let us recall the kinds of problems that these codes were written to solve: smooth, non-stiff, moderately stable problems, for which the function evaluations are moderately to very expensive. It should be realized that our codes have a high overhead and require a lot of storage compared to alternatives. Overhead is important only when solving problems with inexpensive function evaluations and/or not very accurately. In such cases the computation is usually inexpensive in real time and the possibility that methods with lower overhead may be cheaper is unimportant. We have encountered computations in an industrial context which required so many integrations of this kind that real time became significant, and a Runge-Kutta code gave somewhat shorter run times. The codes have seen class use on a PDP-11, a minicomputer. We have seen a problem in an academic environment which could not be contained, along with the differential equation solver, in the core memory of a CDC 6600, a very large computer. Because of the memory transfers which were required, it was slightly cheaper to solve this latter problem using a Runge-Kutta code because, with this method, the whole problem could be held in rapid-access memory. Both of these situations are rare.

The numerical results that follow were all obtained using a CDC 6600, the FUN compiler, single precision, and STEP and INTRP unless stated otherwise. For purposes of cross referencing the examples are numbered.

1. The first example illustrates the starting behavior. The problem is

$$y' = y\left(\frac{4x^3 - y}{x^4 - 1}\right), \quad y(2) = 15.$$

Because the solution is a polynomial of degree 3, $y(x) = 1 + x + x^2 + x^3$, we anticipate that for a reasonably strict error tolerance the code will raise its order to (at least) 3. Because the optimal order is so sharply defined this example illuminates the workings of the order selection algorithm but it should not be regarded as typical. With a pure absolute error test of 10^{-7} the initial step sizes (rounded to three figures) and orders are shown in the following table.

H	K
7.43E-5	1
1.49E-4	2
2.97E-4	3
5.49E-4	4
1.19E-3	5
2.38E-3	4

With no external limit on the step size, it continues to double at every step, indefinitely, and the order stays 4. The code left the start phase when the order was dropped to 4. Why did it overshoot on the order? The solution values are all nearly exact, so the start phase is ended when a signal to lower the order is received. We might have hoped that at order 4 the code would lower the order and leave the start phase at that point. However, the test is whether $\max(\text{ERKM2}, \text{ERKM1}) \leq \text{ERK}$. At order 4 both ERK and ERKM1 are nearly zero, but ERKM2 is substantial since order 2 is not exact. For the same reason, the order stays at 4 after leaving the start phase. The order could be lowered if we were to estimate the error at order 5, since the exactness at orders 3, 4, and 5 might lead to lowering the order by the test $\text{ERKM1} \leq \text{ERK} \leq \text{ERKP1}$. However, the estimate ERKP1 is formed only when the code is working with a constant step size, but here the code is doubling the step size after every step. If we externally restrict the step size to 1.0, the results of continuing the previous integration are as follows.

H	K
2.38E-3	4
⋮	⋮
1.00E0	4
1.00E0	4
1.00E0	4
1.00E0	4
1.00E0	4
1.00E0	3
1.00E0	3

At order 4, five steps of constant step size must be taken before ERKP1 is formed. We see that the test then drops the order to 3. The order does not stay at 3 but fluctuates from 3 to 5 (mostly at 3 and 4) because all three orders are exact (in principle) and roundoff will sometimes reverse inequalities like $\text{ERKM1} \leq \text{ERK} \leq \text{ERKP1}$.

Examples 2 through 5 illustrate the performance on constant coefficient problems. Their format is typical of all the remaining examples. Tolerances of $10^{-2}, 10^{-3}, \ldots, 10^{-13}, 10^{-14}$ are used. For each tolerance the problem is integrated from 0 to about 50, the exact limit depending on the problem. The error measure is part of the problem. The errors reported are, unless stated otherwise, the maximum error found in the specified measure at any step in the integration over the specified interval. Normally errors are reported for the intervals (0, 10), (0, 30), and (0, 50). The cost of the computation is measured by ND, the number of derivative evaluations (the number of calls to the routine for evaluating the right hand side of the differential equation).

At the most stringent error request, $\varepsilon = 10^{-14}$, and sometimes for larger values, the code indicated that too much accuracy was being requested. In all cases the integration was continued using the EPS returned by the code. This situation is indicated by an asterisk next to the tolerance in the tables. The test for limiting precision is intended to be generous so that nearly the best possible accuracy can be obtained. This means that sometimes the gains in accuracy due to reducing ε near limiting precision are not really worth the increased cost. At the other end of the tolerance range, we sometimes see absurd global errors on the long intervals of integration. This arises when the local error control is so crude that the code drifts away from the desired solution and starts tracking a solution curve with very different behavior.

In Chapter 6 we gave the rule of thumb that the maximum global error will be roughly equal to ε; this is amply illustrated in the tables. In Chapter 7 we pointed out that reducing ε by a factor of 0.1 should reduce the global error sufficiently that the accuracy of the less accurate result can be reliably estimated by consistency. Nearly all the tables report the maximum error in any solution component over the entire interval of integration. A study of the tables shows that this maximum decreases, as it must if our contention about the global error is correct. This lends support to the statement that the global errors decrease at each point, but the quantities tabulated do not show it directly. Example 19 is devoted specifically to this question.

2. $y' = y$, $y(0) = 1$, on (0, 10), (0, 30), (0, 50) using relative error test. The solution is $y(x) = \exp(x)$, which grows monotonely, so relative error is appropriate. The CDC 6600 has the unit roundoff $U \doteq 7.1 \times 10^{-15}$. Since the code aims at 0.5ε and keeps this at least as large as $2U \|y_n\|$, we see that for this problem

$$0.5\varepsilon \geq 2 \times 7.1 \times 10^{-15} \times 1$$

on the first call. That is, $\varepsilon \geq 2.82 \times 10^{-14}$ must be true on the first call or the tolerance will be rejected. Thus all $\varepsilon \leq 10^{-14}$ in the table were increased on the first call and the results are all the same. After this example we do not show results for tolerances which are rejected, except for the first one.

LOG$_{10}(\epsilon)$	(0,10) ERROR	ND	(0,30) ERROR	ND	(0,50) ERROR	ND
- 2	2.00E- 2	35	2.16E- 2	85	2.25E- 2	135
- 3	1.40E- 3	49	4.89E- 3	93	7.85E- 3	137
- 4	3.87E- 4	59	4.43E- 4	127	4.76E- 4	197
- 5	2.52E- 5	79	2.91E- 5	187	1.18E- 4	263
- 6	7.15E- 6	89	1.07E- 5	190	1.41E- 5	290
- 7	5.63E- 7	109	1.54E- 6	227	2.38E- 6	351
- 8	3.75E- 8	127	4.85E- 8	278	4.85E- 8	442
- 9	2.94E- 9	155	6.39E- 9	366	9.96E- 9	583
-10	3.09E-10	202	4.97E-10	483	4.97E-10	762
-11	1.97E-11	235	5.30E-11	516	5.30E-11	824
-12	5.71E-12	229	7.16E-12	581	1.71E-11	937
-13	1.65E-12	306	6.32E-12	733	3.04E-11	1176
-14*	1.67E-12	348	2.18E-11	840	4.59E-11	1333
-15*	1.67E-12	348	2.18E-11	840	4.59E-11	1333
-16*	1.67E-12	348	2.18E-11	840	4.59E-11	1333

3. $y' = -y$, $y(0) = 1$, on (0, 10), (0, 30), (0, 50) using relative error test. The solution is $y(x) = \exp(-x)$, which decreases monotonely, so it is not clear that a relative error test is appropriate. If we want to follow a decreasing solution, we must use relative error; later this problem is reconsidered when we are not interested in the solution after it is sufficiently small.

LOG$_{10}(\epsilon)$	(0,10) ERROR	ND	(0,30) ERROR	ND	(0,50) ERROR	ND
- 2	1.45E- 1	46	1.81E- 1	122	1.81E- 1	202
- 3	6.30E- 3	58	1.48E- 2	161	1.67E- 2	267
- 4	2.65E- 4	78	6.42E- 4	200	1.06E- 3	330
- 5	7.64E- 5	92	1.79E- 4	239	2.14E- 4	383
- 6	2.30E- 6	114	2.30E- 6	292	6.06E- 6	471
- 7	5.12E- 7	125	7.23E- 7	320	1.36E- 6	505
- 8	3.67E- 8	152	1.25E- 7	362	1.44E- 7	592
- 9	1.83E- 9	206	2.80E- 9	521	1.32E- 8	777
-10	4.39E-10	214	4.39E-10	513	5.89E-10	812
-11	4.20E-11	248	6.31E-11	624	8.30E-11	994
-12	2.82E-12	283	2.52E-11	714	6.60E-11	1190
-13	1.81E-12	468	1.63E-11	1028	6.83E-11	1619
-14*	1.99E-12	421	1.34E-11	1034	5.70E-11	1656

4. $y_1' = y_2$, $y_2' = -y_1$, $y_1(0) = 0$, $y_2(0) = 1$ on $(0, 2\pi)$, $(0, 6\pi)$, $(0, 16\pi)$ using absolute error test. The solution components are $y_1(x) = \sin x$, $y_2(x) = \cos x$. Because they oscillate between ± 1 and vanish, an absolute error test is appropriate.

	$(0, 2\pi)$		$(0, 6\pi)$		$(0, 16\pi)$	
LOG$_{10}(\epsilon)$	ERROR	ND	ERROR	ND	ERROR	ND
- 2	4.14E- 2	27	8.61E- 2	70	1.50E- 1	177
- 3	3.22E- 3	37	7.04E- 3	90	2.29E- 2	219
- 4	1.21E- 4	53	1.21E- 4	131	4.67E- 4	330
- 5	4.69E- 5	59	9.01E- 5	151	1.67E- 4	365
- 6	1.73E- 6	77	4.84E- 6	175	1.55E- 5	419
- 7	1.87E- 7	105	7.61E- 7	217	1.92E- 6	525
- 8	1.95E- 8	109	6.69E- 8	259	2.19E- 7	648
- 9	4.19E-10	147	1.35E- 9	341	2.38E- 8	746
-10	2.85E-10	161	5.12E-10	362	1.51E- 9	895
-11	6.41E-12	189	2.20E-11	433	6.51E-11	1039
-12	1.26E-12	199	1.21E-11	455	7.07E-11	1100
-13	1.87E-13	275	4.42E-12	614	1.70E-11	1464
-14*	9.67E-13	337	6.24E-12	734	4.08E-11	1722

By manipulating the interval and tolerance on this problem we can exercise the major flow paths in DE. Since the structure of the code is simple it is easy to verify which paths are taken. It is convenient to start the integrations at $x = \pi$ where $y_1(\pi) = 0$, $y_2(\pi) = -1$, and we always do this. In all cases we use pure absolute error, so RELERR = 0.0. The major possibilities and example situations exercising them are:

(i) Improper input:
    There are a number of improper inputs. As an example we solve the problem on $[\pi, 2\pi]$ with input values ABSERR = 0.0 and IFLAG = -1 to terminate internally at 2π. Since both ABSERR and RELERR are zero, DE returns immediately with IFLAG = 6.

(ii) Successful call
    (a) Interpolate: solve on $[\pi, 2\pi]$ with input values ABSERR = 10^{-7}, IFLAG = 1 and then on $[2\pi, 3\pi]$ using the output of the first call. Answers at 2π and 3π are obtained by interpolation and IFLAG = 2.
    (b) Extrapolate: solve on $[\pi, \pi(1 + 2U)]$, where U is the unit round-off, with input values ABSERR = 10^{-7}, IFLAG = 1. The output value is obtained by extrapolation, and IFLAG = 2.

(iii) Unsuccessful call
    (a) Error tolerance too small: solve on $[\pi, 2\pi]$ with input values

ABSERR = U, the unit roundoff, IFLAG = 1. The code does not take a step and returns with IFLAG = 3. Other instances of this return are found in examples 14 and 16.

(b) Too much work: solve on $[\pi, 16\pi]$ with input values ABSERR = 10^{-13}, IFLAG = 1. The code does not get all the way to 16π in 500 steps, and so it returns with IFLAG = 4. Other instances of this return are found in examples 14 and 16.

(c) Stiff problem: this cannot be demonstrated here, since this example is not stiff and the code does not give a false return. Example 14 shows the behavior on a stiff problem.

(iv) Unusual call

(a) Repeated calls with IFLAG negative: solve on $[\pi, 2\pi]$ and then on $[2\pi, 3\pi]$ with ABSERR = 10^{-7}. Input IFLAG = -1 on the first call. Set the output value of IFLAG to -2 and call again for the second interval. This causes DE to restart. The computation is successful and the final output value of IFLAG is 2.

(b) Reversing direction of integration: solve on $[\pi, 2\pi]$ and then on $[2\pi, \pi]$ with ABSERR = 10^{-7}. Input IFLAG = 1 on the first call and use its output value on the second call. This causes DE to restart. The computation is successful and the final output value of IFLAG is 2.

5. $y' = -y$, $y(0) = 1$ on (0, 10), (0, 30), (0, 50) using absolute error test. The solution is $y(x) = \exp(-x)$. The solution eventually drops below the error tolerance and its integration is a question of stability rather than accuracy, as discussed in Chapter 8. The results are

	(0,10)		(0,30)		(0,50)	
LOG$_{10}(\epsilon)$	ERROR	ND	ERROR	ND	ERROR	ND
-2	7.55E-3	31	3.40E-2	58	3.40E-2	80
-3	7.86E-4	44	1.28E-3	69	1.36E-3	100
-4	3.23E-5	57	1.39E-4	81	4.34E-4	99
-5	1.23E-5	67	2.54E-5	108	2.54E-5	127
-6	2.16E-7	89	1.86E-6	122	3.74E-6	144
-7	5.13E-8	103	1.78E-7	143	5.12E-7	176
-8	2.39E-9	123	2.66E-8	170	2.66E-8	190
-9	7.06E-10	157	2.98E-9	213	2.98E-9	243
-10	6.57E-11	169	2.56E-10	241	3.84E-10	270
-11	6.90E-12	205	2.53E-11	293	4.47E-11	320
-12	6.67E-13	213	7.62E-13	324	1.16E-12	345
-13	7.46E-14	269	7.46E-14	389	7.49E-14	419
$-14*$	4.51E-14	308	4.51E-14	464	3.95E-13	489

The integration from 30 to 50 shows how the code responds to mild stiffness. The step size is now limited by stability rather than by accuracy. The step size is rather erratic, as the analysis of Chapter 8 suggests. This is because, if a step gives an unusually small local error, the code will try a larger step size which will work acceptably until the errors due to instability grow sufficiently large to be detected. The code will then reduce both the step size and the order to cope with this growth. Because it is a conservative choice, this step size proves to be too small after the effects of instability are damped out. An effective step size on the interval (30, 50) can be deduced by dividing half of the total number of function calls into the length of the interval (half because a successful step uses two calls). For example, with $\varepsilon = 10^{-2}$ there are 22 calls between $x = 30$ and $x = 50$, an effective average step size of 1.82. Proceeding in this way for all the tolerances we obtain the following table.

$\text{LOG}_{10}(\varepsilon)$	EFFECTIVE STEP SIZE
- 2	1.82
- 3	1.29
- 4	2.22
- 5	2.11
- 6	1.82
- 7	1.21
- 8	2.00
- 9	1.33
-10	1.38
-11	1.48
-12	1.90
-13	1.33
-14*	1.60

The arguments of Chapter 8 state that the effective step size should correspond to being in the region of stability for the order used. In addition the code should be efficient enough so that it works with about as large a step size as it can. Let $-H_k$ be the point at which the boundary of the stability region for the method of order k intersects the negative real axis. For this problem $f_y = -1$; therefore the requirement that hf_y be in the region of stability is simply that $h \le H_k$. The H_k decrease fairly rapidly for the larger k, so we expect that the code would use rather low orders, and it does. Since $H_1 \doteq 2.0$ and $H_2 \doteq 2.4$ we might expect that the order will not drop to 1 and, except for one tolerance, it does not. Most steps over the interval (30, 50) are taken at orders 2 through 4 and in this range of orders $1.4 \le H_k \le 2.4$. Comparing the effective step sizes to these bounds shows that the step used is about as large as is possible, consistent with the stability requirements. The global errors show that stability limitations are not leading to any deterioration of the accuracy, much less catastrophic error growth. This example exhibits quite a mild stability constraint; later we shall examine a very severe constraint.

6. We include one fairly large system, even though its behavior is like that of the preceding example. From a semi-discretization of the heat equation $u_t(t, x) = u_{xx}(t, x)$ a family of problems is obtained, which depends on the number of equations N to be used. The numerical results displayed are for $N = 50$. The initial value problem is

$$y' = Ay,$$

$$A = \begin{pmatrix} -2 & 1 & & & & 0 \\ 1 & -2 & 1 & & & \\ & 1 & -2 & 1 & & \\ & & & \ddots & & \\ 0 & & & & 1 & -2 \end{pmatrix},$$

$$y(0) = (1, 0, \ldots, 0)^T.$$

The orthogonal matrix R of the normalized eigenvectors of A is given by

$$R_{ij} = \sqrt{\frac{2}{N+1}} \sin\left(\frac{ij\pi}{N+1}\right), \qquad 1 \le i, j \le N$$

and the eigenvalues are

$$\lambda_j = -4 \sin^2\left(\frac{j\pi}{2(N+1)}\right), \qquad 1 \le j \le N.$$

Since $R^{-1}AR = \text{diag}\{\lambda_j\}$ the solution is

$$y(x) = R \cdot \text{diag}\{\exp(\lambda_i x)\} \cdot R^{-1} \cdot y(0).$$

The eigenvalues and the explicit solution show that the components decay, so an absolute error test is appropriate. The eigenvalues show that this family is a little more stiff than the preceding example, but the numerical results show that the computed solution is entirely satisfactory.

	(0, 10)		(0, 30)		(0, 50)	
LOG $_{10}(\epsilon)$	ERROR	ND	ERROR	ND	ERROR	ND
- 2	1.69E- 2	69	1.78E- 2	174	1.78E- 2	296
- 3	8.85E- 4	77	2.78E- 3	191	2.78E- 3	291
- 4	9.95E- 5	94	2.12E- 4	189	2.12E- 4	293
- 5	1.19E- 5	112	1.66E- 5	215	1.66E- 5	329
- 6	5.25E- 7	142	1.31E- 6	250	1.42E- 6	377
- 7	2.01E- 7	166	3.28E- 7	288	3.28E- 7	400
- 8	3.61E- 9	198	2.20E- 8	327	2.20E- 8	463
- 9	5.17E-10	233	2.37E- 9	400	2.37E- 9	533
-10	8.49E-11	280	1.40E-10	453	1.40E-10	601
-11	4.36E-12	331	2.85E-11	526	2.85E-11	680
-12	3.54E-13	384	1.14E-12	618	1.26E-12	797
-13	7.70E-14	483	1.24E-13	746	1.24E-13	943
-14*	6.39E-14	615	6.39E-14	900	6.39E-14	1099

Examples 7 through 10 involve nonlinear equations. The equations are simple but give a good indication of how the codes perform on problems of celestial mechanics. They are also used to point out other aspects of the codes.

7.
$$y'_1 = y_2 y_3, \qquad y_1(0) = 0,$$
$$y'_2 = -y_1 y_3, \qquad y_2(0) = 1,$$
$$y'_3 = -0.51 y_1 y_2, \qquad y_3(0) = 1$$

with an absolute error test on $(0, 4K)$, $(0, 12K)$, $(0, 28K)$. These are the Euler equations of motion for a rigid body without external forces. The solution components are the Jacobian elliptic functions $y_1(x) = \text{sn}(x \mid 0.51)$, $y_2(x) = \text{cn}(x \mid 0.51)$, $y_3(x) = \text{dn}(x \mid 0.51)$. They are periodic with a quarter period $K = 1.86264080233273855203\ldots$. The plot shows their general behavior.

	(0, 4K)		(0, 12K)		(0, 28K)	
LOG $_{10}(\epsilon)$	ERROR	ND	ERROR	ND	ERROR	ND
- 2	7.93E- 3	46	4.80E- 2	120	1.06E- 1	274
- 3	2.03E- 3	65	7.30E- 3	159	2.17E- 2	358
- 4	1.98E- 4	81	2.33E- 4	212	5.37E- 4	485
- 5	1.22E- 5	103	5.42E- 5	279	5.88E- 5	648
- 6	1.30E- 6	141	7.01E- 6	376	4.17E- 5	847
- 7	2.04E- 7	155	5.74E- 7	437	6.61E- 6	1007
- 8	6.49E- 9	204	9.77E- 8	526	4.16E- 7	1264
- 9	1.92E- 9	241	6.09E- 9	644	4.42E- 8	1530
-10	6.87E-10	265	4.53E- 9	769	1.65E- 8	1786
-11	1.81E-12	379	1.15E-10	943	8.68E-10	2162
-12	2.31E-12	375	1.74E-11	937	1.10E-10	2059

LOG$_{10}(\epsilon)$	(0,4K) ERROR	ND	(0,12K) ERROR	ND	(0,28K) ERROR	ND
-13	1.62E-12	511	3.47E-12	1294	1.19E-10	3062
-14*	1.74E-12	651	2.12E-11	1792	6.33E-11	4078

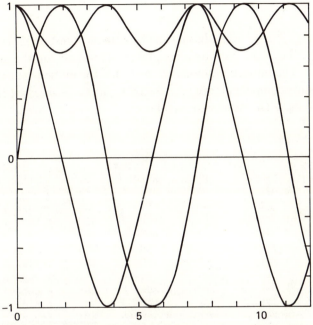

The Jacobian elliptic functions sn($x\,|\,0.51$), cn($x\,|\,0.51$), and dn($x\,|\,0.51$) plotted here are the solution of the Euler equations of motion of a rigid body considered in Example 7.

The restriction of step size increases to a factor of two causes a certain crudity in the adjustment of the step size to the solution, which is only partially compensated for by using an "optimal" factor when reducing the step size. In a particular integration this just means the code may be conservative and a global error obtained which is smaller than would result from a perfect adjustment. However, sometimes reducing the tolerance by a factor of 0.1 results in little change in the global error. This problem exhibits a tendency toward such "ratchetting" of the global error. It shows that caution must be used in assessing the global error by the method of reintegration discussed in Chapter 7. Because of the periodic variation of the solutions in this problem a variation of the step size is appropriate. Since the ideal ratio of the maximum step size to the minimum is about 1.5 for this problem we should expect some tendency to ratchet because of the step size algorithm.

8.

$$y_1'' = -\frac{y_1}{r^3}, \qquad y_1(0) = 1, \qquad y_1'(0) = 0,$$

$$y_2'' = -\frac{y_2}{r^3}, \qquad y_2(0) = 0, \qquad y_2'(0) = 1$$

where $r = (y_1^2 + y_2^2)^{1/2}$ and using an absolute error test on $(0, 2\pi)$, $(0, 6\pi)$, $(0, 16\pi)$. These are Newton's equations of motion for the two body problem. The initial conditions are such that the motion is circular and the solution components are $y_1(x) = \cos x$, $y_1'(x) = -\sin x$, $y_2(x) = \sin x$, $y_2'(x) = \cos x$. The results are as follows.

	(0,2π)		(0,6π)		(0,16π)	
LOG$_{10}(\epsilon)$	ERROR	ND	ERROR	ND	ERROR	ND
- 2	3.96E- 1	29	1.75E 0	74	2.27E 0	219
- 3	2.75E- 2	43	4.35E- 2	104	3.84E- 1	253
- 4	8.34E- 4	61	9.42E- 3	136	1.80E- 1	313
- 5	4.31E- 4	61	2.45E- 3	141	1.85E- 2	352
- 6	4.16E- 5	83	1.28E- 4	201	1.63E- 4	494
- 7	1.25E- 6	101	1.87E- 5	236	1.71E- 4	547
- 8	1.39E- 7	121	1.29E- 6	267	9.36E- 6	633
- 9	1.69E- 8	163	3.96E- 8	395	6.56E- 8	974
-10	1.17E- 9	159	3.12E- 9	365	2.51E- 8	899
-11	3.26E-10	203	8.97E-10	491	1.26E- 9	1214
-12	1.14E-11	213	1.82E-10	512	1.26E- 9	1225
-13	6.93E-13	383	4.94E-12	783	9.83E-11	2112
-14*	7.51E-13	363	1.02E-11	934	8.62E-11	2364

The test on the input error tolerance in STEP causes the request of $\epsilon = 10^{-14}$ to be increased. One might think that without this control the code would return on minimum step size if it is called with ϵ too small. This is not the case. Going below this limit leads to increasingly bad results at a rapidly increasing cost. We suppressed this feature of the code to show what will happen on this problem.

	(0,2π)		(0,6π)		(0,16π)	
LOG$_{10}(\epsilon)$	ERROR	ND	ERROR	ND	ERROR	ND
-14	1.47E-12	641	1.26E-11	1565	8.43E-11	3229
-15	4.54E-12	1084	1.63E-11	3100	4.50E-11	8141
-16	4.30E-12	3369	1.23E-10	9906	9.58E-10	26662

9.

$$y_1'' = -\frac{y_1}{r^3}, \qquad y_1(0) = 0.4, \qquad y_1'(0) = 0,$$

$$y_2'' = -\frac{y_2}{r^3}, \qquad y_2(0) = 0, \qquad y_2'(0) = 2$$

where $r = (y_1^2 + y_2^2)^{1/2}$ and an absolute error test is used on $(0, 2\pi)$, $(0, 6\pi)$, $(0, 16\pi)$. This is similar to example 8 with the exception that the initial conditions cause the motion to be an ellipse with an eccentricity of 0.6. The solution components are $y_1(x) = \cos u - 0.6$, $y_1'(x) = \sin u/(1 - 0.6 \cos u)$, $y_2(x) = 0.8 \sin u$, $y_2'(x) = 0.8 \cos u/(1 - 0.6 \cos u)$ where u is the solution of Kepler's equation $x = u - 0.6 \sin u$. The plot shows the orbits for this and the preceding example.

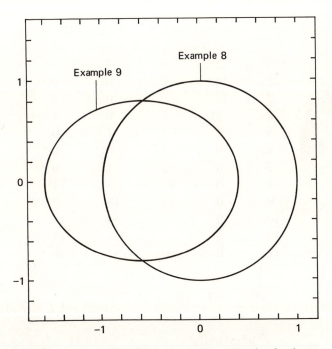

Examples 8 and 9 consider Newton's equations of motion for the two body problem in circular and elliptic motion, respectively.

	(0,2π)		(0,6π)		(0,16π)	
$\log_{10}(\epsilon)$	ERROR	ND	ERROR	ND	ERROR	ND
-2	7.74E- 1	60	2.76E: 0	223	3.33E+ 0	847
-3	2.12E- 2	93	1.29E: 0	278	2.55E ÷ 0	722
-4	1.51E- 3	125	2.49E- 1	347	1.26E ÷ 0	914
-5	5.65E- 5	164	2.03E- 2	441	2.12E- 1	1161
-6	3.99E- 5	193	1.29E- 3	552	1.04E- 2	1414
-7	2.26E- 6	244	7.84E- 5	656	1.15E- 3	1754
-8	3.89E- 7	301	2.65E- 5	840	1.10E- 4	2152
-9	3.44E- 8	358	3.73E- 7	989	5.84E- 6	2536
-10	1.11E- 9	422	4.95E- 8	1148	4.85E- 7	3017
-11	2.77E-12	524	6.65E- 9	1399	1.11E- 7	3558
-12	1.43E-10	629	1.72E- 9	1714	1.90E- 9	4386
-13	2.48E-11	852	1.03E-10	2250	6.60E-10	5666
-14*	3.45E-11	946	1.48E- 9	2535	5.28E- 9	6397

This problem requires quite a lot of step size and order changing. It exhibits a rather typical dependence of the order upon the tolerance. The percentage of steps taken at each order during the integration, for each of several tolerances, has been tabulated as follows.

		ε			
		10^{-1}	10^{-4}	10^{-7}	10^{-10}
Order	1	2	0	0	0
	2	16	0	0	0
	3	75	0	0	0
	4	8	0	0	1
	5	0	1	1	0
	6	0	2	0	0
	7	0	35	1	0
	8	0	45	3	1
	9	0	14	19	0
	10	0	1	34	14
	11	0	0	11	25
	12	0	0	31	58

10.
$$y_1'' = 2y_2' + y_1 - \frac{\mu^*(y_1 + \mu)}{r_1^3} - \frac{\mu(y_1 - \mu^*)}{r_2^3}, \quad y_1(0) = 1.2, \quad y_1'(0) = 0,$$

$$y_2'' = -2y_1' + y_2 - \frac{\mu^* y_2}{r_1^3} - \frac{\mu y_2}{r_2^3}, \quad y_2(0) = 0,$$

$$y_2'(0) = -1.04935750983031990726$$

with an absolute error test on $(0, T)$ where

$$T = 6.19216933131963970674.$$

Here $r_1 = ((y_1 + \mu)^2 + y_2^2)^{1/2}$, $r_2 = ((y_1 - \mu^*)^2 + y_2^2)^{1/2}$, $\mu = 1/82.45$, and $\mu^* = 1 - \mu$. These are the equations of motion of a restricted three body problem, such as a satellite moving under the influence of the earth and the moon in a coordinate system rotating so as to keep the positions of the earth and moon fixed. A high precision computation with these initial conditions gave T as the period of the motion, and the result is accepted as exact. There is no analytic expression available for the solution, hence the error is measured only at the end of the period. This fact must be kept in mind because the errors reported could be much smaller than the worst errors committed during the integration—we do not know one way or the other. To gain more confidence in the accuracy during the course of the integration we can use the Jacobi integral $J(y_1(x), y_2(x), y_1'(x), y_2'(x))$. This function, which is defined by

$$J(y_1, y_2, y_1', y_2') = \frac{1}{2}(y_1'^2 + y_2'^2 - y_1^2 - y_2^2) - \frac{\mu^*}{r_1} - \frac{\mu}{r_2},$$

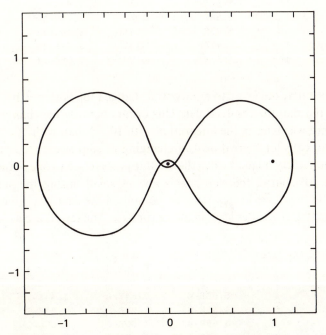

The solution of the restricted three body problem considered in Example 10.

has a constant value along any solution curve. This can be seen by forming dJ/dx and replacing y_1'' and y_2'' with their values from the differential equation to conclude that $dJ/dx = 0$. The true constant value is determined by the initial conditions. We computed the maximum difference between the Jacobi integral of the computed solutions and the true value. This is a valuable test on all components but it must be interpreted carefully. Any discrepancy is due to errors in the computed values but, if the discrepancy is small, we cannot conclude that the values are accurate. They may be accurate or their errors may be correlated so as to give the integral the proper value.

| | | $(0,T)$ | |
| | GLOBAL | | JACOBI INTEGRAL |
$LOG_{10}(\epsilon)$	ERROR	ND	ERROR
- 2	1.19E 0	346	1.31E 0
- 3	9.39E- 3	373	3.55E- 2
- 4	5.91E- 3	516	6.17E- 3
- 5	3.20E- 4	654	3.32E- 4
- 6	1.95E- 5	801	2.10E- 5
- 7	1.86E- 6	959	1.55E- 6
- 8	3.84E- 7	1146	1.57E- 7
- 9	5.30E- 8	1363	1.75E- 8
-10	3.65E- 9	1618	1.52E- 9
-11	1.83E-10	1984	1.06E-10
-12	2.03E-11	2546	1.44E-11
-13*	2.47E-11	3129	9.39E-12
-14*	7.37E-11	3174	2.08E-11

Notice that the error tolerance was changed for both $\epsilon = 10^{-13}$ and 10^{-14} but that the results differ. This occurs because the change of the tolerance was later in the integration with 10^{-13} than with 10^{-14}. This problem calls for a great deal of changing of step size as the problem becomes nearly singular when the satellite passes close to one of the other bodies. Of course, the step size is always small during the start. To illustrate the range in step sizes we computed the smallest, largest, and average step sizes after leaving the start phase. The results are as follows.

$LOG_{10}(\epsilon)$	MIN H	AVG H	MAX H
- 2	1.07E-3	3.86E-2	2.75E-1
- 3	1.07E-3	3.57E-2	2.74E-1
- 4	5.63E-4	2.54E-2	1.44E-1
- 5	4.40E-4	1.98E-2	1.25E-1
- 6	3.65E-4	1.60E-2	1.68E-1
- 7	4.16E-4	1.33E-2	1.06E-1
- 8	3.10E-4	1.11E-2	7.48E-2

$\log_{10}(\epsilon)$	MIN H	AVG H	MAX H
-9	1.74E-4	9.28E-3	5.40E-2
-10	2.75E-5	7.78E-3	5.15E-2
-11	8.69E-6	6.31E-3	3.46E-2
-12	2.75E-6	4.91E-3	2.85E-2
-13	8.69E-7	3.73E-3	2.29E-2
-14	5.85E-7	3.60E-3	2.51E-2

11. To illustrate the effects of precision we solved example 4 on $(0, 16\pi)$ using several machines and word lengths. The machines and unit roundoffs U are

   (i)   IBM 360/67   single precision, 9.5E-7,
   (ii) UNIVAC 1108   single precision, 1.5E-8,
   (iii) PDP-10   single precision, 7.5E-9,
   (iv) CDC 6600   single precision, 7.1E-15,
   (v)   IBM 360/67   double precision, 2.2D-16.

An asterisk beside the reported error indicates that the tolerance had to be increased to complete the integration.

$\log_{10}(\epsilon)$	(i) ERROR	ND	(ii) ERROR	ND	(iii) ERROR	ND
-2	1E-1	177	1E-1	177	1E-1	177
-3	3E-2	218	2E-2	219	2E-2	219
-4	2E-3	304	5E-4	330	5E-4	330
-5	3E-4	363	2E-4	371	2E-4	371
-6	9E-4*	428	3E-5	425	2E-5	425
-7			3E-5	732	4E-6	666
-8			7E-5*	805	8E-6*	715

$\log_{10}(\epsilon)$	(iv) ERROR	ND	(v) ERROR	ND
-2	2E-1	177	1D-1	177
-3	2E-2	219	2D-2	219
-4	5E-4	330	5D-4	330
-5	2E-4	365	2D-4	365
-6	2E-5	419	2D-5	419
-7	2E-6	525	2D-6	525
-8	2E-7	648	2D-7	648
-9	2E-8	746	2D-8	738
-10	2E-9	895	2D-9	893
-11	7E-11	1039	5D-11	1039
-12	7E-11	1100	2D-11	1133
-13	2E-11	1464	2D-12	1408
-14	4E-11*	1722	1D-12	1670
-15			2D-12	1796
-16			4D-15*	1887

Evidently the nature of the machine arithmetic plays an insignificant role until limiting precision is approached. Recall that the point at which the control of propagated roundoff begins depends on the tolerance and the unit roundoff. Near limiting precision the error tends to stop decreasing and the cost begins rising more than it ought to. For each machine and precision, the code is detecting and handling limiting precision very adequately.

12. In Chapter 9 we discussed several simple devices for improving the roundoff characteristics of the codes. One of these attempts to improve the situation with regards to propagated error. As the analysis suggests the effects are small. In some instances the global errors are smaller and in others the cost is reduced. The device is not always useful, this being particularly true of stiff problems when the supporting analysis is invalid. The solutions of the orbit problems (examples 8, 9, and 10) all benefit considerably from its use. We report the results with and without the control for example 9 to show the kind of benefit that may be seen. The tolerances reported are the only ones for which the control is switched on by the code.

	\(0,2\pi)\		\(0,6\pi)\		\(0,16\pi)\	
$LOG_{10}(\epsilon)$	ERROR	ND	ERROR	ND	ERROR	ND
			WITH CONTROL			
-12	1.43E-10	629	1.72E- 9	1714	1.90E- 9	4386
-13	2.48E-11	852	1.03E-10	2250	6.60E-10	5666
-14*	3.45E-11	946	1.48E- 9	2535	5.28E- 9	6397
			WITHOUT CONTROL			
-12	4.76E-10	615	3.67E- 9	1683	1.86E- 8	4361
-13	5.95E-10	822	3.59E- 9	2243	2.18E- 8	5715
-14*	6.28E-10	916	2.03E- 9	2417	1.55E- 8	6316

13. The starting algorithm is intended to find a suitable order and step size very quickly. As discussed in Chapter 7 we expect that the start phase will normally end because the proper order is found rather than because the step size is too large for the current order. We tested this in some of the preceding examples, namely 2 through 5 and 7 through 10, for the tolerances 10^{-2} to 10^{-14}. Very few instances of step failures in the start phase were encountered. They occurred at the tolerances 10^{-2} in

example 7; 10^{-2} and 10^{-3} in example 9; and 10^{-4} in example 10. This accords with the analysis of Chapter 7 which shows that normally such failures will occur only for relatively large tolerances.

Examples 14 through 18 show what happens when the codes are used on problems for which they are not intended. Because of their robust design they perform quite well.

14. The code DE attempts to detect stiffness in certain circumstances. The code always monitors the number of steps and returns control to the user if 500 steps are taken. This usually corresponds to a little more than 1000 derivative evaluations and allows the user to control the work expended in a reasonable way. The test for stiffness is a sequence of at least 50 consecutive steps taken with orders in the range of 1 to 4 and a return because of too much work. This test and its limitations are fully discussed at the end of Chapter 8, and in Chapter 10 there are two illuminating exercises (2 and 3) on this point. The following problem shows the behavior for a typical stiff problem.

The problem is

$$\begin{pmatrix} y_1' \\ y_2' \end{pmatrix} = \begin{pmatrix} 2 - 3E4, & 4 - 4E4 \\ 1.5E4 - 1.5, & 2E4 - 3 \end{pmatrix} \begin{pmatrix} y_1 \\ y_2 \end{pmatrix},$$

$$y_1(0) = 1, \qquad y_2(0) = 1,$$

to be solved with an absolute error test. The solution components are

$$y_1(x) = 7 \exp(-10^4 x) - 6 \exp(-x),$$

$$y_2(x) = -3.5 \exp(-10^4 x) + 4.5 \exp(-x).$$

We integrate this from 0 to 50 using DE. It returns control to the calling program either because of too much work (IFLAG $= 4$) or, for some tolerances, because of too much accuracy having been requested (IFLAG $= 3$). We continue integrating until reaching 50 or until stiffness is indicated by IFLAG $= 5$. We have tabulated, for each tolerance, the point where the integration was terminated (rounded to 3 decimals), the maximum error made at any step during the integration, and the cost in derivative evaluations. As usual, an asterisk indicates that the tolerance was increased during the integration. In every case, we terminated the integration because stiffness was indicated.

LOG$_{10}(\epsilon)$	X	ERROR	ND
- 2	0.163	7.01E- 2	2147
- 3	0.153	7.67E -3	2126
- 4	0.076	6.86E- 4	1072
- 5	0.075	5.03E- 5	1074
- 6	0.078	7.65E- 6	1075
- 7	0.069	4.97E- 7	1063
- 8	0.068	3.53E- 8	1070
- 9	0.064	4.81E- 9	1055
-10	0.141	5.93E-10	2119
-11	0.124	8.99E-11	2078
-12	0.938	2.20E-11	14569
-13*	2.098	2.55E-11	35687
-14*	2.819	3.36E-11	46104

The eigenvalues of the matrix are -1 and -10^4, so that the solution components are linear combinations of $\exp(-x)$ and $\exp(-10^4 x)$. Stiffness simply means that there is a potential for very rapid change, like $\exp(-10^4 x)$, but that the solution of interest changes relatively

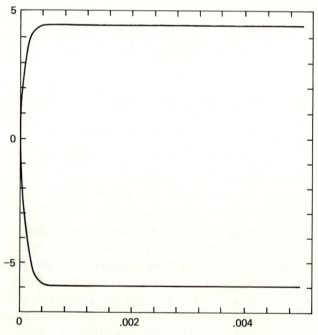

The solution of the stiff problem of Example 14 in a region where the components change very rapidly.

slowly. The plots of the solutions over two different intervals show a region of very rapid change followed by gradual change. The problem is not stiff in this first region but it is expensive to integrate because of the rapid change. Once past this region the problem is stiff and is expensive because the Adams methods are inappropriate, though they will meet the accuracy request. For tolerances of 10^{-2} to 10^{-11}, DE worked rather well at detecting stiffness and controlling the amount of work. Sometimes there is a return of too much work followed by a return indicating stiffness (approximately 2000 evaluations in the table), and other times there is a single return indicating stiffness (approximately 1000 evaluations in the table). Both possibilities are to be expected in view of the problem changing from non-stiff to stiff as the integration proceeds. The propagated roundoff control is applied by the code automatically for the tolerances 10^{-12} to 10^{-14}. The justification for this control, as explained in Chapter 9, relies on the problem being non-stiff, so we do not know what will happen on a stiff problem. There is a marked difference in behavior for these tolerances. The integration

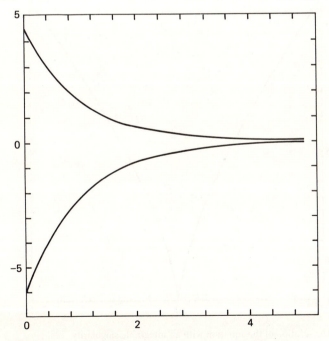

The solution of the stiff problem of Example 14 in a region where the components change slowly.

proceeded very much farther before detecting stiffness and cost a great deal. We cannot predict what will happen at limiting precision, but for the tolerances 10^{-12}, 10^{-13}, and 10^{-14}, DE returned 14, 35, and 46 times, respectively, with the statement that too much work was being expended. Repeated messages to this effect should be taken as being an emphatic warning that the Adams code is inappropriate.

15.

$$y' = \frac{2}{3}x^{-1/3} \qquad \text{if } x \neq 0,$$

$$= 0 \qquad \text{if } x = 0,$$

$y(-1) = 1$ on $(-1, 0)$ and $(-1, 1)$ with an absolute error test. The solution $y(x) = x^{2/3}$. This is an example of an integrable singularity. One really ought to avoid such problems because they require a low order

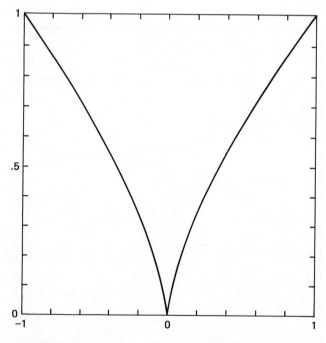

The solution of the equation with an integrable singularity considered in Example 15.

formula and are expensive, but the code *can* integrate them. The results are given in the following table.

LOG$_{10}(\epsilon)$	(-1,0)		(-1,1)	
	ERROR	ND	ERROR	ND
- 2	6.68E- 2	15	6.68E- 2	31
- 3	9.22E- 3	27	3.96E- 2	68
- 4	5.01E- 3	43	5.01E- 3	118
- 5	8.40E- 4	90	8.40E- 4	255
- 6	8.60E- 5	116	2.27E- 4	343
- 7	2.66E- 5	181	2.99E- 5	524
- 8	1.55E- 6	259	1.55E- 6	746
- 9	1.75E- 6	321	2.31E- 6	914
-10	2.94E- 8	416	3.30E- 8	1207
-11	3.13E- 9	559	3.13E- 9	1546
-12	3.15E-10	704	3.15E-10	1939
-13	3.60E-11	878	3.60E-11	2361
-14*	1.55E-11	1006	1.56E-11	2713

16. We cannot say just what will happen when we try to integrate a non-integrable singularity. Using the driver DE we might find that too much work is expended, or that too much accuracy is requested, or, worst of all, that the problem will be integrated. An examination of the problem $y' = y^2$, $y(0) = 1$ on (0, 1) shows what can happen. The solution is $y(x) = 1/(1 - x)$. When we use DE with an absolute error tolerance of 10^{-7} and ask for the value of $y(x)$ at $x = 1$, the code returns at $x = 0.999998341861794$ with the statement that too much work has been expended (IFLAG = 4). If we continue, the code returns at $x = 0.999999883382038$ with a solution value of $y = 3.53608277136011E + 6$ and the statement that too much accuracy was requested (IFLAG = 3). Such a response on the CDC 6600 is very rare for "normal" problems because of the word length. However, the test in the code is that

$$2U\,|y_n| \le 0.5\varepsilon$$

which in this case is

$$|y_n| \le \frac{0.5 \times 10^{-7}}{2 \times 7.1 \times 10^{-15}} \doteq 3.52 \times 10^6$$

so it has performed correctly. The true solution at the x where the code terminated is about $8.575E + 6$; therefore, the answer is very inaccurate. This happens because this problem is very unstable for large y. Since $f_y = 2y$ and y becomes very large, the problem itself is extremely sensitive to small changes. One must keep in mind that the codes attempt to control local errors, and whether or not this controls the global error depends on how stable the problem is. When faced with an unstable problem the code performs in a perfectly satisfactory manner at doing the task it was designed for, but from the users' point of view, it has failed. How badly it "fails" is appreciated when one uses a relative error test to solve the problem. When using this test the code believes that it has solved the problem and gives the answer, at $x = 1$, of $y = 2.24369045795178E + 6$. In view of the extremely unstable nature of the problem this is entirely reasonable and makes it clear that some thought is necessary, even with a good code, for the solution of differential equations.

One of the extensions of DE described in the previous chapter estimates the size of the residual during an integration. When the three computations reported above are made with this extended version of DE, the same answers are obtained. With the first error criterion RES $= 3.82E-7$ on both calls and with the second criterion RES $= 1.07E-6$. This verifies that the code really is controlling the local error as it is supposed to do even though the global error has become very large.

17. Often a problem that is "too easy" proves to be a more severe test of a code than an abnormal problem. The following problem with a mild discontinuity causes difficulty in many codes.

$$y' = 0 \qquad -1 \le x \le 0,$$
$$= x^6 \qquad 0 \le x \le 1,$$

$y(-1) = 0$ on $(-1, 1)$ with absolute error test. The solution is $y(x) = 0$ for $-1 \le x \le 0$ and $y(x) = x^7/7$ for $0 \le x \le 1$. The problem is so easy up to the origin that a code's step size selection algorithm will want to increase the step size as rapidly as possible and so it will step well past the origin. It will then find a large error which may cause the step size to be reduced far too much. As a consequence the code "chatters" near the origin and may prove extremely expensive. STEP is cautious in the step adjustment and performs well.

	(-1,1)	
LOG $_{10}(\epsilon)$	ERROR	ND
- 2	8.73E- 3	22
- 3	3.45E- 3	25
- 4	1.15E- 4	38
- 5	5.20E- 5	44
- 6	7.27E- 7	48
- 7	1.38E- 7	58
- 8	2.01E- 8	98
- 9	9.27E-10	67
-10	1.22E-10	58
-11	8.20E-12	77
-12	1.43E-12	78
-13	1.28E-13	85
-14	1.75E-14	108

The step size chosen by STEP was not permitted to exceed 2. It is usually prudent to place some limit on the step size and this is done in DE. As explained in example 1, the order goes to 3 in the start phase of the integration. Once past the origin the order is raised rapidly to 7 when the code is again exact. For tolerances of 10^{-2} to 10^{-6} there are insufficiently many steps taken between the origin and the endpoint $x = 1$ for the code to reach order seven. For all other error requests the code gets to the proper order.

18.
$$y' = y \qquad 0 \leq x \leq 1,$$
$$= -y \qquad 1 < x \leq 2,$$

$y(0) = 1$ on $(0, 2)$ with absolute error test. The solution $y(x) = \exp(x)$ for $0 \leq x \leq 1$ and $y(x) = \exp(2 - x)$ for $1 \leq x \leq 2$ has a jump in its first derivative of $2e$ at $x = 1$. This is a severe discontinuity. Because the order used by the code rises very rapidly in the interval $(0, 1)$ the code must detect the discontinuity and restart. Furthermore, the discontinuity is so severe that at order one the optimal step size is used to locate the discontinuity accurately enough to enable the code to pass it. This example illustrates the device discussed in Chapter 7 for dealing with such discontinuities and shows it to be remarkably effective at delivering the required accuracy, cheaply.

$$(0,2)$$

LOG$_{10}(\epsilon)$	ERROR	ND
-2	1.68E- 1	34
-3	5.76E- 3	61
-4	3.27E- 5	94
-5	1.93E- 5	121
-6	4.14E- 7	154
-7	8.28E- 8	194
-8	8.39E- 9	226
-9	8.65E- 9	251
-10	9.65E-11	314
-11	4.16E-12	360
-12	1.62E-12	405
$-13*$	4.26E-13	410
$-14*$	3.55E-13	447

Although the code has proved robust on this example, it is best for the user to supply the location of discontinuities by restarting.

We have used software which monitors the statements in a code as it solves a problem and counts the number of times each statement is executed. The software splits logical IFs so that, for example, in the statement "IF(ERKP1.GE.0.5*ERK) GO TO 445" we get counts of the number of times the test is made and of the number of times control is transferred to the statement numbered 445. The software does not split arithmetic IFs in a corresponding way. Many runs with this software show that most of STEP is executed on every problem. The exceptions are the obvious ones like the provisions for failed steps, crashes, and input tolerances that are too small. Example 18 exercises *every* executable statement in STEP at least once. This includes all the branches from logical IFs, with the single exception that in the statement "IF(ERKP1.GE.0.5*ERK) GO TO 445" control is never passed to 445. Besides serving to verify that all parts of the code are functional this example provides a simple test that one has implemented the code correctly.

19. To illustrate the estimation of global errors by reintegration we return to the restricted three body problem of example 10. As we have pointed out, this difficult problem requires a great deal of step size and order changing. This furnishes a severe test of the arguments of Chapter 7 which state that we can estimate global errors by consistency. We report the estimated and true global errors at the end of one period for the two solution components y_1 and y_2. The estimated global error at a tolerance

ε is obtained by subtracting the result at this tolerance from that for the tolerance 0.1ε. For $\varepsilon = 10^{-13}$ the tolerance is increased during the integration, hence the estimated error corresponding to $\varepsilon = 10^{-12}$ may be unreliable due to its closeness to limiting precision.

The estimated errors are remarkably good, especially since the problem is a severe test of the theory, and this may cause the inexperienced to be unduly optimistic. The result for $y_1(T)$ with $\varepsilon = 10^{-3}$, where the sign is incorrect, emphasizes the need for cautious interpretation of numerical results. Still, these results generally support our contention that such estimates of the global error are likely to be correct to within a factor of two and quite likely to be correct to within an order of magnitude.

	$y_1(T) = 1.2$		$y_2(T) = 0.0$	
$\text{LOG}_{10}(\varepsilon)$	EST. ERROR	TRUE ERROR	EST. ERROR	TRUE ERROR
- 2	8.9E- 1	8.9E- 1	-8.6E- 2	-9.2E- 2
- 3	1.7E- 3	-9.1E- 5	-6.4E- 3	-6.0E- 3
- 4	-2.1E- 3	-1.8E- 3	6.1E- 4	4.1E- 4
- 5	3.3E- 4	3.2E- 4	-1.7E- 4	-1.9E- 4
- 6	-1.1E- 5	-1.3E- 5	-1.9E- 5	-2.0E- 5
- 7	-2.0E- 6	-1.9E- 6	-2.0E- 7	-5.8E- 7
- 8	2.1E- 7	2.0E- 7	-4.0E- 7	-3.8E- 7
- 9	-1.4E- 8	-1.6E- 8	1.2E- 8	1.2E- 8
-10	-1.1E- 9	-1.1E- 9	-2.1E-10	-3.2E-10
-11	-2.3E-11	-2.5E-12	-1.0E-10	-1.1E-10
-12*	1.0E-11	2.0E-11	-1.0E-11	-1.1E-11

12

TECHNIQUES, EXAMPLES, AND EXERCISES

Solving real problems by numerical integration usually involves more than just calling an integrator. There is no substitute for understanding the physical problem and the integrator, and using common sense for arriving at reliable, meaningful results. Often a problem must be reformulated in order to integrate it at all, and in other cases a reformulation may prove more satisfactory. In this chapter we discuss and show by example some of the techniques for effectively using the codes. The reader should regard most of the examples as informal exercises. Enough details and numerical results are supplied that he can program the problem and check his computation, thereby increasing his understanding.

ERROR CONTROL

Since so few users really understand what the codes attempt to do, it is worthwhile to review the matter. When presented with a problem

$$\mathbf{y}' = \mathbf{f}(x, \mathbf{y}), \qquad \mathbf{y}(a) = \mathbf{A}$$

the codes attempt to produce a function $\mathbf{y}_I(x)$ such that

$$\mathbf{y}_I'(x) = \mathbf{f}(x, \mathbf{y}_I(x)) + \mathbf{r}(x), \quad \mathbf{y}_I(a) = \mathbf{A}$$

and $\mathbf{r}(x)$ is "small." The size of the residual, $\mathbf{r}(x)$, depends upon the local error tolerance imposed by the user and whether the codes are successful. Almost always, when the codes report success, they have produced a solution $\mathbf{y}_I(x)$ with a small residual. What this means in terms of the global error

$$\mathbf{y}(b) - \mathbf{y}_I(b)$$

depends on the problem itself.

Ordinarily the global error is about the size of the local error tolerance given to the codes. It is possible for the global error to decrease as well as increase, but generally one must assume a gradual deterioration of the accuracy as the integration proceeds. If the user reduces the tolerance and integrates the problem again, the error of the less accurate result can be reliably estimated by consistency, using the codes provided with this text.

Example 16 of the preceding chapter illustrates very clearly the distinction between control of local error and control of global error. The reader should understand this example and since it is an easy task, should try doing it.

EFFICIENT EVALUATION OF PARTIALLY LINEAR EQUATIONS

By understanding what the codes do and with a little care in the programming, one can sometimes significantly reduce the cost of solving a differential equation. The matter is clearest when solving a linear system of equations

$$\mathbf{y}' = A(x)\mathbf{y} + \mathbf{g}(x) = \mathbf{f}(x, \mathbf{y}).$$

Nearly all the work in evaluating $\mathbf{f}(x, \mathbf{y})$ takes place when the matrix $A(x)$ and the vector $\mathbf{g}(x)$ are evaluated. The Adams codes evaluate $\mathbf{f}$ twice in each step at the same argument x_{n+1}. To take advantage of this fact, just program the subroutine to remember A and $\mathbf{g}$ from the previous evaluation.

Use a variable in COMMON with the main program to tell the subroutine it is being called for the first time. On this first call, evaluate and store x, $A(x)$, and $\mathbf{g}(x)$. Thereafter, every time the subroutine is called it should test the argument x against the x stored. If they are the same, use the stored values of $A(x)$ and $\mathbf{g}(x)$. If they are different, evaluate and store x, $A(x)$, and $\mathbf{g}(x)$.

For a linear system of equations this simple device nearly halves the work of evaluating the equations. Of course the idea is applicable to any set of equations in which some function depending only on x appears. As a very simple example we show how to integrate a forced Duffing's equation from 0 to 10 in such a way as to halve the number of trigonometric evaluations. It is worth noting that COMMON is just one way of handling the initialization problem. The problem is

$$y'' + y - \frac{y^3}{6} = 2 \sin 2.78535x,$$

$$y(0) = 0, \qquad y'(0) = 0.$$

```
      EXTERNAL F
      COMMON XSAVE
      DIMENSION Y(2)
      Y(1)=0.0
      Y(2)=0.0
      T=0.0
      XSAVE=-1.0
      TOUT=10.0
      RELERR=0.0
      ABSERR=1.0E-5
      IFLAG=1
      CALL DE(F,2,Y,T,TOUT,RELERR,ABSERR,IFLAG)
      TYPE 100,Y(1),Y(2),IFLAG
100   FORMAT(1P2E15.7,I15)
      STOP
      END
      SUBROUTINE F(X,Y,YP)
      COMMON XSAVE
      DIMENSION Y(2),YP(2)
      IF(X.NE.XSAVE)TERM=2.0*SIN(2.78535*X)
      YP(1)=Y(2)
      YP(2)=(Y(1)**3)/6.0-Y(1)+TERM
      XSAVE=X
      RETURN
      END
```

LINEAR INDEPENDENCE AND BOUNDARY VALUE PROBLEMS

The theory of linear differential equations is much better developed than that of nonlinear equations and is correspondingly more illuminating of the numerical procedures. Linear differential equations with n components have the form

$$\mathbf{y}' = A(x)\mathbf{y} + \mathbf{g}(x), \tag{1}$$

where $A(x)$ is an $n \times n$ matrix and $\mathbf{y}$ and $\mathbf{g}$ are vectors of n components. As usual we assume that all the functions are continuous. The system is said to be homogeneous if $\mathbf{g}(x) \equiv 0$, and inhomogeneous otherwise. It is known that any solution $\mathbf{y}(x)$ of the homogeneous system

$$\mathbf{y}' = A(x)\mathbf{y} \tag{2}$$

can be represented as a linear combination of n linearly independent solutions $\mathbf{u}^i(x)$:

$$\mathbf{y}(x) = \sum_{i=1}^{n} \alpha_i \mathbf{u}^i(x).$$

A set of functions $\{\mathbf{u}^i(x)\}$ is said to be linearly independent if none of the functions can be represented as a linear combination of the others, i.e., a relation such as

$$\mathbf{0} \equiv \sum_{i=1}^{n} \beta_i \mathbf{u}^i(x) \tag{3}$$

means that all the β_i are zero. It is known that the solutions $\mathbf{u}^1(x), \ldots, \mathbf{u}^n(x)$ of (2) are linearly independent if they are linearly independent at any point x_0. Thus if we were to define the $\mathbf{u}^i(x)$ as the solutions of (2) with the initial conditions

$$u^i_j(x_0) = 1 \qquad \text{if } i = j,$$
$$= 0 \qquad \text{otherwise,}$$

they must be linearly independent. Otherwise, if relation (3) held, then for each component j we would have

$$0 = \sum_{i=1}^{n} \beta_i u^i_j(x_0) = \beta_j.$$

It can be seen that any solution $\mathbf{y}(x)$ of (1) can be written as the sum of any one particular solution $\mathbf{y}^p(x)$ and a solution of (2). Thus

$$\mathbf{y}(x) = \mathbf{y}^p(x) + \sum_{i=1}^{n} \alpha_i \mathbf{u}^i(x).$$

Airey's equation,

$$y'' - xy = 0,$$

is an important equation of mathematical physics. All solutions are a linear combination of any two linearly independent solutions. Two linearly independent solutions which are a standard choice are designated by $Ai(x)$ and $Bi(x)$ and are defined by the initial conditions

$$Ai(0) = \frac{Bi(0)}{\sqrt{3}} = \frac{3^{-2/3}}{\Gamma(\frac{2}{3})} \doteq 0.355028053887817,$$

$$-Ai'(0) = \frac{Bi'(0)}{\sqrt{3}} = \frac{3^{-1/3}}{\Gamma(\frac{1}{3})} \doteq 0.255819403792807.$$

($\Gamma(x)$ is the gamma function). For positive x, $Ai(x)$ decreases exponentially and $Bi(x)$ increases exponentially. Suppose we attempt to compute $Ai(x)$ numerically. By the time we reach $x_n > 0$ we will not have exactly $y_n = Ai(x_n)$, or $y'_n = Ai'(x_n)$. Suppose we write the numerical values that we do have in the form

$$y_n = \alpha Ai(x_n) + \beta Bi(x_n), \qquad y'_n = \alpha Ai'(x_n) + \beta Bi'(x_n)$$

where $\alpha \doteq 1$ and $\beta \doteq 0$. This means that the local solution is

$$u_n(x) = \alpha Ai(x) + \beta Bi(x).$$

Even if the code does a perfect job of following the local solution, as it is intended to do, we find that the approximate solution at x_{n+1} is

$$\alpha Ai(x_{n+1}) + \beta Bi(x_{n+1}).$$

The point is that $Bi(x)$ grows very rapidly compared to $Ai(x)$, with the consequence that the numerical result at x_{n+1} is considerably less accurate than that at x_n. The problem is not very well posed.

If we did not already know how the solutions behaved, we could get some insight by approximating the problem near x_n by

$$y'' - x_n y = 0.$$

This constant coefficient problem is easily solved and has the two linearly independent solutions

$$\exp(x\sqrt{x_n}) \qquad \text{and} \qquad \exp(-x\sqrt{x_n}).$$

Clearly, trying to compute a rapidly decreasing solution will be extremely difficult in view of the likelihood that the code will begin to follow a rapidly increasing solution instead.

How far it is possible to integrate a sub-dominant solution depends on how much it is dominated, and on the code and machine being used. Eventually a code based on controlling local errors will begin tracking a multiple of the dominant solution. For the case at hand one can integrate $Ai(x)$ for a distance with considerable accuracy; but, the larger that x is, the more $Bi(x)$ dominates. Even for a rather modest x one finds that a multiple of $Bi(x)$ is being computed. Obviously it is very easy to integrate any multiple of $Bi(x)$; indeed, we cannot avoid it! This example makes it clear that how difficult a problem is numerically depends not just on the equation but also upon the solution being computed, i.e., upon the initial conditions.

It is easy to verify that the Wronski determinant

$$W(x) = \begin{vmatrix} Ai(x) & Bi(x) \\ Ai'(x) & Bi'(x) \end{vmatrix}$$

is a constant because $W'(x) \equiv 0$. From the initial conditions $W(x) \equiv 1/\pi$. If $Ai(x)$ and $Bi(x)$ are computed and the Wronski determinant numerically evaluated, the solutions can be seen to deteriorate as the integration proceeds.

This text is concerned with the initial value problem which specifies a solution by supplying initial values at some point a. Boundary value problems are also fairly common. Such problems specify a solution by giving some solution values at one point and others at another point. (Still more general situations also arise.) Thus for a boundary value problem we must solve

$$\mathbf{y}' = \mathbf{f}(x, \mathbf{y}), \tag{4a}$$

$$y_{i_m}(a) = A_{i_m} \qquad m = 1, 2, \ldots, p, \tag{4b}$$

$$y_{i_m}(b) = A_{i_m} \qquad m = p + 1, \ldots, n. \tag{4c}$$

The situation with respect to the existence and uniqueness of solutions of this problem is very complex. Depending on the problem there may be no solution, any finite number of distinct solutions, or even an infinite number of solutions. Reference [5] has an introductory chapter with a number of illuminating examples. If the equations are linear, the matter is comparatively simple. For a given interval $[a, b]$ there is either exactly one solution for any set of boundary values, or else there are boundary values for which there is no solution and other values for which there are infinitely many solutions.

For linear problems the solution of a well posed boundary value problem is, in principle, quite simple. Using a code for the initial value problem we can find a particular solution $\mathbf{y}^P(x)$ of the inhomogeneous equations (1) and a set of linearly independent solutions $\mathbf{u}^1(x), \ldots, \mathbf{u}^n(x)$ for the homogeneous problem (2). Then any solution $\mathbf{y}(x)$ of the equations (1) can be written

$$\mathbf{y}(x) = \mathbf{y}^P(x) + \sum_{j=1}^n \alpha_j \mathbf{u}^j(x).$$

Thus, for $\mathbf{y}(x)$ to be the solution of the boundary value problem (4), we must have

$$A_{i_m} - y_{i_m}^P(a) = \sum_{j=1}^n \alpha_j u_{i_m}^j(a) \qquad m = 1, 2, \ldots, p,$$

$$A_{i_m} - y_{i_m}^P(b) = \sum_{j=1}^n \alpha_j u_{i_m}^j(b) \qquad m = p + 1, \ldots, n.$$

This is a set of n linear algebraic equations for the n unknowns α_j. If there is a unique solution, the boundary value problem has a unique solution. Otherwise there are some sets of boundary values $\{A_{i_m}\}$ for which the boundary value problem has no solution, and others for which there are infinitely many; in either case the problem is not well posed.

Whether or not this approach is feasible depends on how easily a set of linearly independent solutions can be computed. This is fundamentally a question of how well posed the *initial* value problem is. Unfortunately, it is common that the boundary value problem will have a solution quite insensitive to the given values while, at the same time, the initial value problem has solutions that are very sensitive to the initial values. Sometimes it makes a great deal of difference in which direction the initial value problem is integrated. Still, one should not be unduly discouraged. Codes for the initial value problem are far more advanced than those aimed directly at boundary value problems. It is relatively easy to try initial value techniques, and they are often successful. There are a number of important devices for improving the performance of initial value techniques on the unstable problems which may arise in this way [8, 14]

The virtue of the approach outlined for linear problems is that one can almost immediately solve the boundary value problem for any set of boundary values after first constructing the fundamental solutions. It is more work than is necessary if only a single problem is to be solved. By incorporating known values (4b) at $x = a$ it is not necessary to compute certain of the $\mathbf{u}^j(x)$. To be specific, consider the problem

$$y''(x) + (3 \cot x + 2 \tan x)y'(x) + 0.7y(x) = 0$$

$$y(30) = 0, \qquad y(60) = 5.$$

(The angles are measured in degrees.) This problem arises when considering the stress distribution in a spherical membrane having normal and tangential loads. The general approach might define

$$u^1(30) = 1, \qquad \frac{d}{dx}u^1(30) = 0,$$

$$u^2(30) = 0, \qquad \frac{d}{dx}u^2(30) = 1,$$

$$y^p(30) = 1, \qquad \frac{d}{dx}y^p(30) = 0.$$

However, if we choose $y^p(x)$ so as to satisfy one boundary condition, for instance,

$$y^p(30) = 0, \qquad \frac{d}{dx}y^p(30) = 1, \tag{5}$$

then clearly, if

$$y(x) = y^p(x) + \alpha_1 u^1(x) + \alpha_2 u^2(x),$$

the boundary condition at $x = 30$ implies $\alpha_1 = 0$. Thus, we need only integrate the inhomogeneous equation (which in this example is homogeneous) with the initial conditions (5), and then integrate the homogeneous equation once with

$$u^2(30) = 0, \qquad \frac{d}{dx}u^2(30) = 1.$$

The constant α_2 is determined from

$$y(60) = 5 = y^p(60) + \alpha_2 u^2(60).$$

Of course this particular example is homogeneous, so $y^p(x)$ is simply a multiple of $u^2(x)$ and it is only necessary to do a single integration.

In the general case we can reduce the work by a considerable amount with this device, which is a sketch of the method of complementary functions [23]. Integration would normally be started at the end where the larger number of values is specified. However, it is very important to integrate in a direction for which the initial value problem is stable. An important reason for taking advantage of known boundary values is that they restrict the kinds of solutions which are possible for the initial value problem. In our example we do not need to compute $u^1(x)$. If $u^2(x)$ is easy to compute and $u^1(x)$ is difficult, the method of complementary functions will succeed and the general approach may fail.

For this example the variable coefficient can be approximated by its average value 5.2458729. If the reader solves the constant coefficient boundary value problem analytically, he should be able to see why it is feasible to integrate from 30 to 60 but not in the reverse direction. The unknown initial slope at $x = 30$ is computed to be about 1901, and computational results show that integrating from 60 to 30 is unsatisfactory.

Nonlinear problems are much more complex. One approach resembles the method of complementary functions. This method integrates or "shoots" from one end point to the other using guessed values for all the unknown initial conditions. A nonlinear equation solver is used to adjust the unknown initial values at this end point so that the computed solution agrees with the known values at the other end point. Another approach is to approximate the nonlinear problem by a sequence of linear ones and to solve the linear problems by the methods already discussed. The reader interested in non-linear boundary value problems should consult references [5, 14, and 22].

There are a number of interesting problems which can be converted to initial value problems by using groups of transformations [15]. An example is the Blasius problem

$$\frac{d^3r}{dx^3} + r\frac{d^2r}{dx^2} = 0,$$

$$r(0) = 0, \qquad \frac{dr}{dx}(0) = 0, \qquad \lim_{x \to \infty} \frac{dr(x)}{dx} = 1.$$

The solution, or rather its derivative $r'(x)$, gives a solution to the laminar flow of an incompressible fluid with small viscosity moving parallel to a semi-infinite flat plate. Many computational solutions of this and related problems have guessed a value for $r''(0) = s$ and computed the solution $r(x; s)$ of the equation with initial values

$$r(0) = 0, \qquad r'(0) = 0, \qquad r''(0) = s.$$

Then an attempt is made to find that value of s such that at some "large" value b

$$r'(b; s) = 1.$$

Using the codes DE and ROOT this is quite easy to do. Having taken $b = 20$ we found s to be about 0.4696. For this particular problem we can proceed differently. It is known that the derivative of the solution of the initial value problem

$$w^{(3)}(z) + w(z)w^{(2)}(z) = 0,$$

$$w(0) = 0, \qquad w'(0) = 0, \qquad w^{(2)}(0) = 1$$

increases to a limit λ^2,

$$\lim_{z \to \infty} w'(z) = \lambda^2 > 0.$$

The reader can verify that the solution of the Blasius problem is given in terms of the solution of this initial value problem and the limit value λ as

$$r(x) = \frac{1}{\lambda} w(z) \qquad \text{for } z = \frac{x}{\lambda}.$$

Thus we have

$$\frac{d^2 r(0)}{dx^2} = \frac{1}{\lambda^3} \left(\frac{d^2 w(0)}{dz^2} \right).$$

Using this approach and integrating to $z = 20$ to determine the limit, we found s to be about 0.4696, in agreement with the more general approach.

ROOT SOLVING

The statistical distributions of Karl Pearson are defined as the solutions of

$$y' = -y \left(\frac{c_1 + x}{c_0 + c_1 x + c_2 x^2} \right) \qquad L_1 < x < L_2, \tag{6}$$

where the coefficients c_0, c_1, c_2 depend on moment ratios β_1, β_2 observed in experiments. The coefficients are:

$$r = \frac{-6(\beta_2 - \beta_1 - 1)}{2\beta_2 - 3\beta_1 - 6}, \qquad \omega = \beta_1(r + 2)^2 + 16(r + 1)$$

$$c_0 = \frac{r + 1}{r - 2}, \qquad c_1 = \frac{\sqrt{\beta_1}(r + 2)}{2(r - 2)}, \qquad c_2 = \frac{-1}{r - 2}.$$

The various types of distributions are characterized by the signs of r and ω; Type IV, of interest to us because they are usually evaluated by solving (6), has $r < 0$, $\omega < 0$. A normalizing condition

$$\int_{L_1}^{L_2} y(x)\, dx = 1 \tag{7}$$

is imposed on the solution of (6) to uniquely specify it.

Frequently statisticians are not so much interested in the distribution function itself as they are in the cumulative distribution

$$F(x) = \int_{L_1}^{x} y(t)\, dt \qquad L_1 \le x \le L_2. \tag{8}$$

In particular they are concerned with the "percent points" of F, i.e., those points x_ρ satisfying $F(x_\rho) = \rho$. To determine these points for Type IV distributions, statisticians integrate the system (6) and (8) along with the normalization condition (7) and the initial value $F(L_1) = 0$.

Here we see still another way to specify a solution of a differential equation. We must reformulate the problem as an initial value problem so as to be able to use our codes. Equation (6) is linear and homogeneous, so any solution $v(x)$ is a multiple of the solution $y(x)$ that we are seeking, i.e.,

$$y(x) = \alpha v(x).$$

We impose arbitrary initial conditions, for example, $v(0) = 0.5$, $G(0) = 0$ and integrate

$$v'(x) = -v\left(\frac{c_1 + x}{c_0 + c_1 x + c_2 x^2}\right), \qquad v(0) = 0.5,$$

$$G'(x) = v(x), \qquad G(0) = 0$$

first on $[0, L_1]$ and then $[0, L_2]$. The normalizing factor, α, is determined by

$$\int_{L_1}^{L_2} y(x)\, dx = \alpha \int_{L_1}^{L_2} v(x)\, dx = \alpha \int_{L_1}^{0} v(x)\, dx + \alpha \int_{0}^{L_2} v(x)\, dx = 1$$

from which we see that

$$\alpha = \frac{1}{G(L_2) - G(L_1)}.$$

The solution $y(x)$ then becomes known in terms of α and the computed solution $v(x)$. The cumulative distribution $F(x)$ follows from its definition:

$$F(x) = \int_{L_1}^{x} y(t)\, dt = \alpha \int_{L_1}^{0} v(t)\, dt + \alpha \int_{0}^{x} v(t)\, dt$$

$$= -\alpha G(L_1) + \alpha G(x).$$

Storing the computed solutions is troublesome so the typical procedure is to repeat the integration with initial conditions

$$y(0) = \alpha v(0) = \alpha \cdot 0.5,$$

$$F(0) = -\alpha G(L_1) + \alpha G(0) = -\alpha G(L_1),$$

to determine $y(x)$, $F(x)$ and the percent points.

The exact solution $y(x)$ may be written

$$y(x) = y(0) \cdot \frac{\exp\left\{ v \arctan\left(\dfrac{x}{a} - \dfrac{x}{r}\right)\right\}}{\left\{1 + \left(\dfrac{x}{a} - \dfrac{x}{r}\right)^2\right\}^m}$$

where

$$a = \frac{\sqrt{-\omega}}{4}, \qquad v = \frac{\sqrt{\beta_1} r(r+2)}{\sqrt{-\omega}}, \qquad m = \frac{2-r}{2}.$$

Although the distribution can be found analytically, the cumulative distribution cannot, so the problem must be solved numerically. The exact solution for $y(x)$ provides a check on how well the integrator is performing.

The interval of integration for a Type IV distribution is infinite, $L_1 = -\infty$, $L_2 = \infty$. How do we know where to end the integration? Superficially, the answer is simple: quit when the results would not be significantly altered if the integration were continued. A rigorous justification requires an analysis of how rapidly the solution dies off, and error bounds are usually impossible to obtain. The distribution $y(x)$ is a skewed bell curve, and from the exact solution we see that the tails decrease exponentially at infinity and so have little effect on the solution. Statisticians normally terminate the integration when

$$|y(x)| < \varepsilon$$

where ε is the error tolerance of the integrator. This criterion is also practical since the solution decreases steadily below this point and the integrator cannot follow it. The value $G(x)$ when the integration is terminated is accepted as $G(L_1)$ or $G(L_2)$.

It is quite common in solving differential equations to seek the value of the independent variable for which a dependent variable has a given value. In this particular problem we want the value of x such that $F(x) = \rho$. One way to do this is to define a new system in which F is the independent variable and x is a dependent variable, and then integrate from a known value of F to ρ. The initial conditions for the new system are derived from the old system. The value of the dependent variable x at the independent variable ρ is the percent point x_ρ.

From equation (8) we see that

$$\frac{dx}{dF} = \frac{1}{y}$$

so the new system, with F as independent variable, is

$$\frac{dy}{dF} = \frac{dy}{dx} \cdot \frac{dx}{dF} = -\frac{c_1 + x}{c_0 + c_1 x + c_2 x^2},$$

$$\frac{dx}{dF} = \frac{1}{y}. \tag{9}$$

From the known values for $x = 0$, y_0 and F_0, the initial values are found to be

$$x(F_0) = 0, \qquad y(F_0) = y_0.$$

To make up a table of percent points we merely need to call DE repeatedly to integrate to the various $F = \rho$ which interest us, and the corresponding $x(\rho)$ values are the desired percent points.

This simple treatment is possible because $F(x)$ is a monotonely increasing function for all x, which allows us to introduce it as the independent variable without having to take any further care. In other problems dF/dx might vanish somewhere so that the change of variables cannot be made blindly. To deal with such problems one should step along using STEP until $F(x_n) < \rho < F(x_{n+1})$. The change of variables is then introduced at x_{n+1} and the short distance to ρ is integrated. The trouble with this general approach is that for each ρ the equations must be changed, and accordingly it is necessary to restart in order to obtain each percent point.

Another approach to finding the percent point x_ρ is to use a root solver for nonlinear equations. Several good ones are available in the literature; in Chapter 10 we supply one that is well adapted to use with differential equation solvers. Again we step along using STEP until $F(x_n) < \rho < F(x_{n+1})$. The code ROOT requires an interval $[B, C]$ which brackets the root to be found. Here we solve $F(x) - \rho = 0$ with $B = x_n$, $C = x_{n+1}$. The root solver must be able to evaluate the equation anywhere in this interval. This is easily done using the interpolation code INTRP. It is important to appreciate that no evaluations of the differential equation are required to solve the task posed, only evaluations of an interpolating polynomial.

The Type IV distribution specified by $\beta_1 = 1.44$ and $\beta_2 = 6.0$ is a skewed bell curve. With $v(0) = 0.5$, $G(0) = 0$ we integrated the initial value problem using double precision on an IBM 360/67, an absolute local error tolerance of 10^{-11}, and STEP. The integrations terminated at $L_1 = -3.16$ and $L_2 = 35.8$ at which points the solutions had fallen below the tolerance: $G(L_1) = -0.687$ and $G(L_2) = 0.520$ leading to $\alpha = 0.828$. The following percent points of the normalized curve were computed by both methods and agree with tabulated results.

ρ	x_ρ
.05	−1.305
.50	−0.161
.75	0.518
.95	1.847

It is noteworthy that $F(x)$ is quite flat for large x, with the consequence that percent points for ρ near either 0 or 1 are poorly determined. Since $F \to 1-$ and $dF/dx \to 0+$ as $x \to +\infty$ we realize that $dx/dF \to +\infty$ as $F \to 1-$. Thus for ρ near 1 the change of variables means that we integrate close to a singular point; the same happens for ρ near 0. For ρ sufficiently near 0 or 1 it is better to use the root solver than the change of variables. This is because $F(x)$ "looks" like a polynomial there (it approaches a constant) but $x(F)$ does not, and the codes are based on approximation by polynomials.

TRANSFORMING THE INDEPENDENT VARIABLES

For some problems, derivatives become infinite or very large in the original variables with the result that a change of variables may be necessary. Other problems are easier to integrate numerically or are more naturally described in different variables. We have just examined a problem where a change of variables was a convenient way to obtain roots of certain equations, and we also saw that a large derivative in one variable may be easy to handle in another.

There are two general approaches that are used in dealing with inconveniently large derivatives. Suppose we wish to solve

$$\frac{dy_1}{dx} = f_1(x, y_1, \ldots, y_n),$$

$$\vdots \qquad \vdots$$

$$\frac{dy_n}{dx} = f_n(x, y_1, \ldots, y_n).$$

If at x_m we find that

$$1 < \left|\frac{dy_j}{dx}\right| = \max_{1 \le i \le n} \left|\frac{dy_i}{dx}\right| = \max_{1 \le i \le n} |f_i(x_m, y_1, \ldots, y_n)|,$$

we can take y_j to be the new independent variable and so solve

$$\frac{dy_1}{dy_j} = \frac{\dfrac{dy_1}{dx}}{\dfrac{dy_j}{dx}} = \frac{f_1(x, y_1, \ldots, y_n)}{f_j(x, y_1, \ldots, y_n)},$$

$$\vdots \qquad \vdots$$

$$\frac{dx}{dy_j} = \frac{1}{f_j(x, y_1, \ldots, y_n)},$$

$$\vdots \qquad \vdots$$

$$\frac{dy_n}{dy_j} = \frac{f_n(x, y_1, \ldots, y_n)}{f_j(x, y_1, \ldots, y_n)}.$$

Then at y_j every one of the first derivatives is bounded in magnitude by 1:

$$\left|\frac{dy_i}{dy_j}\right| = \left|\frac{f_i}{f_j}\right| \leq 1 \qquad i \neq j,$$

$$\left|\frac{dx}{dy_j}\right| = \left|\frac{1}{f_j}\right| \leq 1.$$

Recall the example of the non-integrable singularity of the preceding chapter, $dy/dx = y^2$, $y(0) = 1$. If we wish to compute the solution accurately as we approach the singularity, we ought to change variables. We might integrate to some point x_m where

$$\left|\frac{dy}{dx}\right| = |y_m^2| > 1$$

and from that point on solve the equation,

$$\frac{dx}{dy} = \frac{1}{y^2}, \qquad x(y_m) = x_m,$$

instead. In the original variables smaller and smaller steps must be taken as the singularity at 1 is approached. This could result in the code reaching its machine-dependent minimum step size. A more fundamental reason for the change is that the solution does not behave like a polynomial and its inverse function does. In this particular example the problem is very unstable for large y, whereas the transformed problem is stable.

A disadvantage of simply interchanging variables is that a numerical method with memory, such as those embodied in the codes of this text, must be restarted each time the switch is made. In some circumstances many switches will be necessary. There may also be the inconvenience of having to use a root solving code if one wishes to stop at a given value of x. An alternative is to use the arc length, s, as the independent variable for the entire integration. In this variable the general equation becomes

$$\frac{dy_1}{ds} = \frac{f_1}{S},$$

$$\vdots$$

$$\frac{dy_n}{ds} = \frac{f_n}{S},$$

$$\frac{dx}{ds} = \frac{1}{S}, \qquad x(a) = 0$$

where

$$S = \left(1 + \sum_{i=1}^{n} f_i^2\right)^{\frac{1}{2}}.$$

Using this variable a root solver will surely be needed in order to be able to stop at a given value of x. There is an additional equation to integrate but this is unimportant. Noting that

$$\left(\frac{dx}{ds}\right)^2 + \sum_{i=1}^{n} \left(\frac{dy_i}{ds}\right)^2 = \frac{1}{S^2} + \sum_{i=1}^{n} \left(\frac{f_i}{S}\right)^2 \equiv 1$$

we realize that

$$\left|\frac{dy_i}{ds}\right| \leq 1 \qquad i = 1, 2, \ldots, n,$$

$$\left|\frac{dx}{ds}\right| \leq 1$$

for the *entire* interval in s. If some dy_j/dx is much larger than the rest, we can see that $dy_j/ds \doteq 1$, and that in effect y_j is introduced locally as a new independent variable. Thus there is a close relation between the two apparently very different approaches.

The example of the non-integrable singularity is

$$\frac{dy}{ds} = \frac{y^2}{\sqrt{1 + y^4}}, \qquad y(0) = 1,$$

$$\frac{dx}{ds} = \frac{1}{\sqrt{1 + y^4}}, \qquad x(0) = 0.$$

To exploit all the power of the change of variables technique, the subroutine for evaluating the equation should be programmed with some care. The difficulty, as seen in this example, is that if y^2 is large, then y^4 may overflow. This accidental overflow can be avoided by evaluating the f_i, $F = \max |f_i|$, and then

$$\frac{f_j}{S} = \frac{\left(\dfrac{f_j}{F}\right)}{\left[\left(\dfrac{1}{F}\right)^2 + \displaystyle\sum_{i=1}^{n} \left(\dfrac{f_i}{F}\right)^2\right]^{\frac{1}{2}}}.$$

There may very well be harmless underflows and the setting of some quantities to zero, but overflows are avoided.

As an example of a problem best studied in a different variable we consider the voltage e across the discharge tube in the circuit illustrated below. Here, an electric current, at a voltage E, flows through resistance R and charges capacitor C, which is connected across the tube's electrodes. The equation for the variation of the voltage, e, across the discharge tube, as a function of time is

$$\frac{de}{dt} = \frac{E - e}{CR} - \frac{e}{Cr}, \qquad e(0) = 0.$$

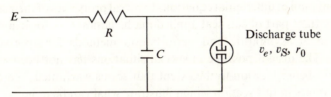

An electrical circuit with an impressed voltage E, a resistance R, capacitance C, and a gas discharge tube.

Here r is the resistance of the discharge tube. As e increases to V_S, the resistance, r, across the tube is infinite. When $e = V_S$ the tube discharges and r changes to the constant value r_0, and e starts to decrease. When e drops below V_e, r again becomes infinite and e starts to increase again. (We require that $E > V_S > r_0 E/(R + r_0)$.) The variation in voltage settles into an oscillation and so we might ask, what is its period? This example is not really a differential equation since the value of the derivative depends on the history of the solution. Consider what r is if $V_e < e(t) < V_S$; it is not definable without knowing how the present value of e was obtained.

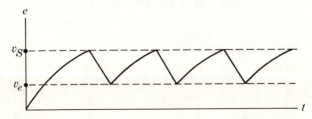

The voltage E across the gas discharge tube oscillates periodically.

In an appropriate variable this problem is easy to analyze and solve. Clearly we should take e to be the independent variable. When we integrate $t(e)$ from 0 to V_S we should take r to be infinite. We should then integrate $t(e)$ from V_S to V_e with $r = r_0$. If we integrate again from V_e to V_S with r infinite, we will have completed the first period of the oscillation. In this way one can use three calls to DE to compute the period. If a table of the solution is wanted, it can be easily generated. For $V_S = 8$, $V_e = 2$, $E = 11$, $CR = 10$, $Cr_0 = 1$ the period is $-10\{\ln 1/3 - (\ln 7)/11\} \doteq 12.75513$.

TRANSFORMING DEPENDENT VARIABLES

Transformation of differential equations to reveal properties of their solutions is an important part of classical applied mathematics. It is, for example, the foundation of asymptotic and perturbation methods for approximating solutions. The insight possible in special situations has not been exploited in numerical computation to the extent that seems warranted. We consider a few examples in this section which illustrate what can be done.

Many interesting physical phenomena are described by equations of the form

$$y'' + v^2 y + \varepsilon f(x, y, y') = 0.$$

Here ε is a small parameter describing the departure of the system from linearity. When $\varepsilon = 0$ the solutions are of the form

$$y(x) = a \sin(vx + \phi)$$

where the amplitude a and the phase ϕ are constant. If we seek to represent the solution for $\varepsilon \neq 0$ in the form

$$y(x) = a(x) \sin(vx + \phi(x)),$$
$$y'(x) = a(x)v \cos(vx + \phi(x)),$$

we readily find that

$$a' = -\frac{\varepsilon}{v} f(x, y, y') \cos(vx + \phi),$$

$$\phi' = \frac{\varepsilon}{av} f(x, y, y') \sin(vx + \phi).$$

Because of the factors of ε appearing in these equations we see that $a(x)$ and $\phi(x)$ are slowly varying for "small" ε. Indeed, all their derivatives are $O(\varepsilon)$. This fact is the basis for a number of analytical techniques for approximating $a(x)$ and $\phi(x)$ for small ε.

These observations are relevant to numerical as well as perturbation methods. The step size is limited by the frequency v of the oscillation even for $\varepsilon = 0$ and the integration over many periods can be expensive. Example 4 of Chapter 11 integrates $y'' + y = 0$ over eight periods and shows that direct integration is already moderately expensive. When $\varepsilon = 0$, the equations for the amplitude and phase can be integrated exactly because the solutions are constant, and in these variables such problems are trivial. The point is that the numerical scheme is based on polynomial approximation and the amplitude and phase "look" a lot more like polynomials than do the original variables. It is clear that advantages are to be gained for small ε by changing variables.

As a simple example we integrated the motion of a spring with a cubic nonlinearity

$$y'' + y - \varepsilon y^3 = 0, \qquad y(0) = 1, \qquad y'(0) = 0.$$

This system is conservative, so the solutions satisfy the energy integral

$$\frac{1}{2}(y^2(x) + y'^2(x)) - \frac{\varepsilon}{4}y^4(x) \equiv \frac{1}{2} - \frac{\varepsilon}{4}.$$

For $\varepsilon = 10^{-4}$ the total energy is 0.499975 which we use to assess the accuracy of the numerical solution. The period of the oscillation is about 2π and we integrated to 100π on a PDP-10 with an absolute error tolerance of 10^{-5}. In the variables y, y', the energy at 100π was computed to be 0.49949119 and the integration required 2189 function evaluations. Using the variables a, ϕ the energy at 100π was 0.49985453 at a cost of 1160 evaluations. This is typical of results over different intervals; better values of the energy are obtained at about half the cost.

Examples 8 and 9 of the preceding chapter are concerned with the motion of two bodies attracting one another with a gravitational force. In suitable units and coordinate system the equations are

$$y_1'' = \frac{-y_1}{r^3}, \qquad y_2'' = \frac{-y_2}{r^3},$$

$$r = \sqrt{y_1^2 + y_2^2}.$$

As the bodies approach, so that y_1 and y_2 both become small, the equations become nearly singular. In general the equations of motion of n bodies acted upon by their gravitational attraction are singular if any pair of bodies collides and nearly singular for close approaches. Astronomers have given a great deal of attention to transforming the equations so as to remove the singular behavior, "regularizing" the equations, and so making them easier to solve numerically. The article by D. G. Bettis and V. Szebehely, "Treatment of close approaches in the numerical integration of the gravitational problem of N bodies," *Astrophysics and Space Science*, 14, 1971, pp. 133–150, is a good introduction to the work being done along these lines. More advanced results pertinent to the viewpoint of this chapter are to be found in the paper by J. Baumgarte and E. Stiefel, "Examples of transformations improving the numerical accuracy of the integration of differential equations," pp. 207–236 of reference [18].

A simple way to regularize the two body equations is to go to the new dependent variables u_1, u_2 by Levi-Civita's transformation

$$y_1 = u_1^2 - u_2^2, \qquad y_2 = 2u_1 u_2, \qquad r = u_1^2 + u_2^2,$$

and to a new independent variable s,

$$\frac{dx}{ds} = u_1^2 + u_2^2.$$

The resulting equations are

$$\frac{d^2 u_1}{ds^2} + \frac{h}{2} u_1 = 0, \qquad \frac{d^2 u_2}{ds^2} + \frac{h}{2} u_2 = 0$$

where $(-h)$ is the constant Kepler energy of the motion, defined by

$$-h = \frac{1}{2}(y_1'^2 + y_2'^2) - \frac{1}{r},$$

and is to be obtained from the initial conditions. The more eccentric the orbit, the more advantage will be gained from this transformation. The disadvantages are that (1) an additional equation is introduced relating the two time variables x and s; (2) there is a cost associated with transforming back to the variables y_1, y_2; and, (3) there is a cost associated with finding the proper value of s so as to terminate the integration at a specified value of x. These disadvantages prove to be minor compared to the improvements in accuracy and cost to be obtained.

We repeated example 9 of Chapter 11 on a PDP-10 with an absolute error tolerance of 10^{-8} and examined the accuracy of the four solution components at 16π. The worst error was 2×10^{-3}, and 2952 function evaluations were required. Integrating with the regularized equations we obtained values for y_1, y'_1, y_2, y'_2 at $x = 16\pi$ which had no error worse than 8×10^{-6} at a cost of only 901 function evaluations. The less eccentric orbit of example 8 in Chapter 11 is less difficult to solve and the gains are less impressive. The reader might enjoy experimenting with this regularization procedure.

The two body problem has a known solution, so the point of these transformations is not to solve this problem easily but instead to solve perturbed two body problems easily. It is reasonable to expect that altering the equations slightly to account for other forces would not alter the conclusion that the regularized equations are easier to solve, and this seems to be the case in practice.

An important task in applied mathematics is to obtain eigenvalues and eigenfunctions of the Sturm-Liouville equation

$$(p(x)u)' + (\lambda\rho(x) - q(x))u = 0, \qquad x\in [a, b]$$

along with the (separated) boundary conditions

$$u(a)\cos A - p(a)u'(a)\sin A = 0,$$

$$u(b)\cos B - p(b)u'(b)\sin B = 0.$$

(A and B are given numbers with $-\pi/2 \le A < \pi/2$, and $-\pi/2 \le B < \pi/2$.) For any value of the parameter λ, the function $u(x) \equiv 0$ is a solution of this problem. For some values of λ, called the eigenvalues, there is a non-trivial solution $u(x)$, called an "eigenfunction" corresponding to the eigenvalue λ. Since any multiple of an eigenfunction obviously satisfies the problem, eigenfunctions are only determined to within a constant factor. If $p(x)\in C^1[a, b]$, $\rho(x)$ and $q(x)$ belong to $C[a, b]$, and both $p(x)$ and $\rho(x)$ are positive, the problem is said to be regular. It is known [reference 7, pp. 263–264] that there is an infinite sequence of real eigenvalues $\lambda_0 < \lambda_1 < \cdots$ with $\lim_{n\to\infty} \lambda_n = \infty$. The eigenfunction $u_n(x)$ corresponding to the eigenvalue λ_n has exactly n zeros in the interval $a < x < b$.

One way to solve the eigenvalue problem is to choose a value λ, initial values $u(a)$ and $u'(a)$ (not both zero) such that

$$u(a)\cos A - p(a)u'(a)\sin A = 0,$$

and then to integrate the resulting initial value problem for its solution $u(x; \lambda)$. This is done repeatedly, using a root-solving code to find a λ such that

$$u(b; \lambda) \cos B - p(b)u'(b; \lambda) \sin B = 0.$$

Such a λ is an eigenvalue and $u(x; \lambda)$ is the corresponding eigenfunction. This is called a "shooting method" and, in particular, we have described shooting from a to b. Obviously it is possible to shoot from b to a and there are still other variants.

There are advantages to using phase and amplitude variables. One possibility is the Prüfer transformation

$$u(x) = r(x) \sin \theta(x), \qquad p(x)u'(x) = r(x) \cos \theta(x).$$

This leads to

$$\theta' = \frac{\cos^2 \theta}{p(x)} + (\lambda p(x) - q(x)) \sin^2 \theta,$$

$$\theta(a) = A, \qquad \theta(b) = B + n\pi$$

where n is a non-negative integer. This says that an eigenvalue may be computed using only the phase variable. After determining the eigenvalue, an eigenfunction is computed by integrating the equation for $\theta(x)$ again, along with

$$r' = \left[\frac{1}{p(x)} - (\lambda p(x) - q(x)) \right] \frac{r \sin 2\theta}{2}.$$

We have stated a result which says that the eigenfunctions associated with the higher eigenvalues must oscillate rapidly. For reasons that we have already explored it seems clear that the phase and amplitude variables are increasingly superior to the original variables as one goes to the higher eigenvalues. This observation has not received the emphasis it deserves in the literature, which has focussed on the convenience of the change of variables for dealing with singular points.

A classical substitution enables one to transform the regular Sturm-Liouville system to the simpler form

$$u'' + (\lambda - w(x))u = 0,$$

$$\alpha u(a) + \alpha' u'(a) = 0, \qquad \beta u(b) + \beta' u'(b) = 0.$$

Assuming that $w(x) \in C^1[a, b]$ and that $\lambda > w(x)$ on $[a, b]$ it is possible to determine the behavior of the eigenvalues and eigenfunctions. The result is that if $\alpha'\beta' \neq 0$, then the eigenvalues λ_n are given as $n \to \infty$ by

$$\sqrt{\lambda_n} = \frac{n\pi}{b - a} + O\left(\frac{1}{n}\right),$$

and the eigenfunctions by

$$u_n(x) = \sqrt{\frac{2}{b - a}} \cos \frac{n\pi(x - a)}{b - a} + O\left(\frac{1}{n}\right).$$

(Here the multiple of the eigenfunction has been chosen so that $\int_a^b u_n^2(x)\,dx = 1$.) This oscillatory behavior is unsuitable for codes based on polynomial approximation. If the modified Prüfer transformation

$$u(x) = \frac{R(x)\cos\phi(x)}{\sqrt[4]{\lambda - w(x)}}, \qquad u'(x) = R(x)\sqrt[4]{\lambda - w(x)}\sin\phi(x),$$

is used, the equations become

$$\phi' = -\sqrt{\lambda - w(x)} + \frac{w'(x)}{4(\lambda - w(x))}\sin 2\phi,$$

$$R' = -\frac{w'(x)R\cos 2\phi}{4(\lambda - w(x))}.$$

The behavior of these quantities is known, too. For all sufficiently large λ the solutions of these equations are

$$\phi(x; \lambda) = \phi(a; \lambda) - \sqrt{\lambda}(x - a) + O\left(\frac{1}{\sqrt{\lambda}}\right),$$

$$R(x; \lambda) = R(a; \lambda) + O\left(\frac{1}{\lambda}\right);$$

meaning that, asymptotically, the amplitude is constant and the phase is linear. Thus, as λ increases to large values, codes based on polynomial approximation will become especially effective, rather than especially ineffective as for the original variables.

Using either the Prüfer or modified Prüfer transformations and the codes DE and ROOT that we supply, it is quite easy to write a code to solve the regular Sturm-Liouville problem efficiently and effectively. Many physical problems involve singular points but they pose no special difficulty, as we show in the next section. A reader contemplating the solution of problems involving singular points should look at the papers of Bailey [4] and Banks and Kurowski [6].

SINGULAR POINTS

A singular point of a differential equation is a point where the existence and uniqueness theorems of Chapter 1 do not apply. They are quite common in physical problems, occurring most often when coefficients become infinite at some point or when an inital condition is given at infinity. It is surprising to the inexperienced that singular points, as they arise physically, are often points where the solution is particularly easy to approximate and (with a little special attention) cause no numerical trouble at all. Those topics of the theory of differential equations required to properly account for the singular points are a part of standard courses on applied mathematics and mathematical physics. We consider several representative examples next and show how one might proceed in solving them.

Emden modelled the thermal behavior of a spherical cloud of gas acting under the mutual attraction of its molecules and the classical laws of thermodynamics by

$$y'' + \frac{2}{r}y' + y^{\gamma} = 0,$$

$$y(0) = 1, \quad y'(0) = 0.$$

The coefficient $2/r$ is infinite at $r = 0$; thus, the origin is a singular point. The partial differential equation for the gravitational potential has been reduced to an equation in the radial variable r because of spherical symmetry. The condition $y'(0) = 0$ simply results from symmetry at the origin. There is no physical reason to expect anything other than a smooth solution everywhere. Of course if we can deal with the neighborhood of the origin, for other r the problem is a standard one and its solution is routine.

The basic idea for treating an isolated singularity is to use analytical techniques to approximate the solution near it, and to use standard numerical techniques elsewhere. The Taylor series approach is the simplest and most common. Suppose, on physical or other grounds, that there is a smooth solution at the origin. Then

$$y(r) = y(0) + ry'(0) + \frac{r^2}{2}y''(0) + \cdots.$$

Now $y''(0)$ can be deduced from the equation by letting $r \to 0$. Note that we must use

$$\lim_{r \to 0} \frac{y'(r)}{r} = y''(0),$$

so that

$$y''(0) + 2y''(0) + y^\gamma(0) = 0$$

implies

$$y''(0) = -\frac{1}{3}.$$

Thus

$$y(r) = 1 - \frac{r^2}{6} + \cdots.$$

By using a Taylor series method we could take a step of length h away from the origin, and from then on use our code. This is quite unnecessary for this problem since the code (in effect) uses a Taylor series method to start. The only difficulty is in properly writing the subroutine for evaluating the equation. If we use $y_1 = y$, $y_2 = y'$, the equation is

$$y_1' = y_2, \qquad\qquad y_1(0) = 1,$$

$$y_2' = -\left(\frac{2}{r}\right)y_2 - y_1^\gamma, \qquad y_2(0) = 0.$$

The only thing we must be careful about is that y_2' be evaluated accurately at and near the origin. The solution being flat, the only trouble spot is at the origin itself where we must supply the proper value. For example, we might write

```
SUBROUTINE F(R,Y,YP)
DIMENSION Y(2),YP(2)
COMMON GAMMA
YP(1) = Y(2)
YP(2) = -1.0/3.0
IF (R.GT.0.0) YP(2) = -2.0*Y(2)/R - Y(1)**GAMMA
RETURN
END
```

Notice that we are passing the parameter γ through COMMON so that we can use whatever value is appropriate.

Astrophysicists use the solutions of this problem to estimate the density and internal temperature of stars. They are most concerned with the solutions from the origin to the first zero so that the root solving techniques we have already discussed are useful here, too. There are a couple of known solutions, one of which is for $\gamma = 5$

$$y(r) = \frac{1}{\left(1 + \dfrac{r^2}{3}\right)^{\frac{1}{2}}}.$$

The solution in this case does not vanish, so it is not physically useful, but the reader should solve it numerically for $y(r)$ and $y'(r)$ at several values for r, equal perhaps to $0.5, 1.0, \ldots, 5.0$ and then compare the numerical and exact values.

The situation at the singular point is clearest when the equation is linear because there is a large body of theory concerning solutions of linear differential equations at singular points; for example, see reference [24]. Latzko's equation,

$$\frac{d}{dz}\left\{(1 - z^7)\frac{dy}{dz}\right\} + \lambda z^7 y = 0,$$

$$y(0) = 0, \qquad y(1) \text{ finite},$$

is an eigenvalue problem arising in the theory of heat flow in an eddying fluid. The quantity λ is a parameter in the equation. Obviously $y(x) \equiv 0$ is always a solution of this problem. A value of λ such that the problem has a non-trivial solution is called an eigenvalue and the corresponding solution an eigenfunction. Since the equation is linear and homogeneous, any multiple of a solution is also a solution, so an eigenfunction is only specified to within a constant multiple. Thus we might require, for example, that $y'(0) = 1$. For each value of λ the problem

$$\frac{d}{dz}\left\{(1 - z^7)\frac{dy}{dz}\right\} + \lambda z^7 y = 0$$

$$y(0) = 0, \qquad y'(0) = 1$$

has a solution $y(z; \lambda)$. The parameter λ is an eigenvalue if the boundary condition at $z = 1$ is satisfied. In this way a succession of initial value problems can be solved by varying λ until an eigenfunction is discovered. A root finding code is used to solve the equation $y(1; \lambda)$ finite and so find the eigenvalues. Each time the root finding code requires a value of $y(1; \lambda)$ corresponding to a specific λ, an initial value problem must be solved. Here we integrate or "shoot" from $z = 0$ to $z = 1$. We could as well shoot from $z = 1$ to $z = 0$. Often one direction is more practical than the other, and this is true of Latzko's equation.

If we write out the equation, we see that

$$y'' - \left(\frac{7z^6}{1 - z^7}\right) y' + \left(\frac{\lambda z^7}{1 - z^7}\right) y = 0.$$

The equation has the form

$$y'' + \left(\frac{P(z)}{z - 1}\right) y' + \left(\frac{Q(z)}{(z - 1)^2}\right) y = 0,$$

where $P(z)$ and $Q(z)$ are regular at 1, considered as functions of a complex variable. This is an example of an equation with a regular singular point at $z = 1$ and, as shown in appendix A of reference [24], because $P(1) = 1$ and $Q(1) = 0$, there are two kinds of solutions:

$$y_1(z) = a_0 + a_1(z - 1) + a_2(z - 1)^2 + \cdots$$

and

$$y_2(z) = y_1(z)\beta \ln (z - 1) + b_0 + b_1(z - 1) + \cdots.$$

One kind of solution is finite at $z = 1$ and the other is logarithmically infinite. The boundary condition requires a solution of the first kind. Obviously if we try to shoot from $z = 0$ to $z = 1$ we shall begin tracking a solution which diverges, and the numerical results will be useless. (It is generally true that in dealing with singularities one must integrate away from them to take advantage of analytical results about the behavior of the solution at these points.)

If we substitute the form of the solution $y_1(z)$ into the equation, match coefficients of the powers of $(z - 1)$, and take $y(1) = 1$, we find that

$$y(z) = 1 + \frac{\lambda}{7}(z - 1) + a_2(z - 1)^2 + \cdots.$$

Alternatively, if we simply presumed that a Taylor series existed, we could just evaluate derivatives to get this result. Latzko's equation can be put into standard form using either the usual variables $y_1 = y$, $y_2 = y'$, or the more natural ones $y_1 = y$, $y_2 = (1 - z^7)y'$. Let us discuss the latter. In this case

$$y_1' = \frac{y_2}{(1 - z^7)}, \qquad y_1(1) = 1,$$

$$y_2' = -\lambda z^7 y_1, \qquad y_2(1) = 0.$$

The only difficulty lies in evaluating the equation at the initial point $z = 1$. Because

$$y_1'(z) = \frac{y_2(z)}{1 - z^7} = y'(z) = \frac{\lambda}{7} + 2a_2(z - 1) + \cdots,$$

we simply need to define

$$y_1'(1) = \frac{\lambda}{7}$$

in the routine. Since $y(z)$ and $y'(z)$ are well approximated by polynomials, the first step will move well away from $z = 1$ and thereafter the routine will have no difficulty evaluating the derivatives.

Try computing the smallest eigenvalues of Latzko's equation. They are about 8.727, 152.4, and 435.1. We give a sample code for computing an eigenvalue between 400 and 500.

```
      EXTERNAL F
      DIMENSION Y(2)
      COMMON TLAMDA
      B=400.0
      C=500.0
      KFLAG=1
1     CALL ROOT(TLAMDA,FTLAM,B,C,1.0E-5,1.0E-5,KFLAG)
      IF(KFLAG.GT.0)GO TO 2
      Y(1)=1.0
      Y(2)=0.0
      T=1.0
      TOUT=0.0
      RELERR=1.0E-5
      ABSERR=1.0E-5
      IFLAG=1
      CALL DE(F,2,Y,T,TOUT,RELERR,ABSERR,IFLAG)
      IF(IFLAG.NE.2)GO TO 3
      FTLAM=Y(1)
      GO TO 1
```

```
2              TYPE 100,KFLAG,B
100            FORMAT(I15,1PE15.7)
               STOP
3              TYPE 100,IFLAG
               STOP
               END
               SUBROUTINE F(Z,Y,YP)
               DIMENSION Y(2),YP(2)
               COMMON TLAMDA
               Z7=Z♦♦7
               YP(1)=TLAMDA/7.0
               IF(Z.LT.1.0)YP(1)=Y(2)/(1.0-Z7)
               YP(2)=-TLAMDA♦Z7♦Y(1)
               RETURN
               END
```

A famous problem, first studied by Euler, is to solve

$$\frac{dy}{dz} = y - \frac{1}{z}, \qquad y(+\infty) = 0.$$

When dealing with a condition at infinity it is usual to transfer it to the origin by a change of variables. If we let $x = 1/z$, it is easy to verify that

$$\frac{dy}{dx} = \frac{1}{x} - \frac{y}{x^2}, \qquad y(0) = 0.$$

If we assume that a Taylor series exists for the solution and proceed to evaluate its terms, we obtain

$$\sum_{n=0}^{\infty} (-1)^n n! \, x^{n+1}.$$

But this series has a zero radius of convergence! However, it is still useful since it is an example of an asymptotic series. In this case one can prove that

$$\left| y(x) - \sum_{n=0}^{m} (-1)^n n! \, x^{n+1} \right| \le (m+1)! \, x^{m+2},$$

which means that the error due to terminating the series at the mth term is bounded by the first term dropped. With a convergent series we fix x and let $m \to \infty$ in order to approximate $y(x)$ arbitrarily well. This is not true of asymptotic series like this one where we must fix m and then say that as $x \to 0$, we approximate $y(x)$ arbitrarily well. For a given x we can approximate $y(x)$ only to a limited accuracy. For example, if $x = 0.1$, the magnitudes of the terms are

m	term
8	$8! \times 10^{-9} \doteq 4.0320 \times 10^{-5}$
9	$9! \times 10^{-10} \doteq 3.6288 \times 10^{-5}$
10	$10! \times 10^{-11} \doteq 3.6288 \times 10^{-5}$
11	$11! \times 10^{-12} \doteq 3.9917 \times 10^{-5}$

The terms with $m < 8$ or $m > 11$ are all larger, so the best accuracy that can be obtained at $x = 0.1$ using this series comes from $m = 8$ or 9; an accuracy of about 3.6×10^{-5}.

Our previous examples did not require us to actually step away from the origin but we must do so here. Indeed we must move away some distance in order to be able to integrate the problem at all. The reason is that the problem is stiff near the origin. In Chapter 8 we found that the step size h_{n+1} must satisfy a bound depending on the variation of the equation with respect to y in order for the computation to be stable (see also examples 5 and 14 of Chapter 11). In this case, at order one, we must have something like

$$-2 < h_{n+1} f_y(x_n) = -\frac{h_{n+1}}{x_n^2}$$

holding at x_n, if the computation is to be stable. Obviously we must step away from the origin to have a chance of solving this problem with a code aimed at non-stiff problems.

If we use the asymptotic series with $m = 9$ to approximate $y(x)$ through $x = 0.1$, we can then integrate the equation numerically. The value at 0.1 is known to be accurate to only 3.6×10^{-5} but since the equation is stable we know errors will be damped out. With an absolute error tolerance of 10^{-5} we may expect accuracies of the same order. At $x = 1.0$ we found $y(1) \doteq 0.596347$. There is an expression for the solution

$$y(x) = x \int_0^\infty \frac{\exp(-t)\,dt}{1 + xt}$$

which is readily evaluated by a Laguerre quadrature formula. Using a 40 point formula we confirmed the value for $y(1)$ computed by solving the differential equation.

In summary, analytical techniques are used to determine how the solution behaves at singular points. If it is reasonable to approximate this solution by a Taylor series at the singularity, we need only program the routine for evaluating the equation carefully, so that proper values are returned, and then solve the problem routinely. If this is not reasonable, or if for some other reason it is not feasible to use the code near the singular point, one must step away from the singular point by analytical means, and then the problem can be solved routinely. We have given some typical examples. For further examples we recommend the readable paper of P. B. Bailey [4]. A number of problems of physical origin involving singularities are carefully analyzed and solved with a code for the initial value problem.

A FINAL PROBLEM

As a final example we consider the pull-out problem for synchronous motors. The numerical solution of this problem requires several of the techniques discussed above, so it serves to test the reader's understanding of this chapter. It also supports our earlier assertion that often there is more to solving a real problem than just calling an integrator. The pull-out problem for a synchronous motor supposes that up to the time $t = 0$ the motor is operating at a constant speed under a torque L_0 and at $t = 0$ the additional torque L_1 is applied and held constant so that the total torque is a constant $L_0 + L_1$. The problem is to determine the largest torque L_1 which can be applied without causing the rotor to reach an unstable equilibrium position and probably cause the motor to fail to operate properly. In Stoker's [27, pp. 70–80] solution of the problem a differential equation with a number of special features is solved. The equation is

$$\frac{dv}{dx} = \frac{-\lambda v - \sin x + \gamma}{v}$$

where λ and γ are constants which are exemplified by 0.022 and 0.8 respectively. There is a critical value x_c where

$$\sin x_c = \gamma, \qquad 0 < x_c < \frac{\pi}{2}$$

which, in our case, is arcsin 0.8. There is a saddle point singularity at $x_S = \pi - x_c$, and the integration is to start there.

At x_S, v is to have the value 0. Because $\sin x_S = \gamma$ we see that

$$\frac{dv}{dx} = \frac{0}{0},$$

so this is a singular point of the equation. If we attempt a Taylor series solution at x_S,

$$v(x) = a_0 + a_1(x - x_S) + a_2(x - x_S)^2 + \cdots,$$

we have $a_0 = 0$ because $v(x_S) = 0$. Using L'Hospital's rule

$$a_1 = \lim_{x \to x_S} \frac{dv}{dx} = \frac{\frac{d}{dx}(-\lambda v - \sin x + \gamma)|_{x_S}}{\frac{d}{dx} v|_{x_S}}$$

$$= -\frac{\lambda a_1 + \cos x_S}{a_1}$$

so

$$a_1^2 + \lambda a_1 + \cos x_S = 0.$$

Therefore

$$a_1 = -\frac{\lambda}{2} \pm \sqrt{\frac{\lambda^2}{4} - \cos x_S}$$

but, because $\cos x_S < 0$ we see that there are two solutions for a_1. This is in agreement with Stoker's derivation of the equation which says that there are two solutions of the equation which vanish at x_S. A sketch of the solution to be computed in his approach is shown here.

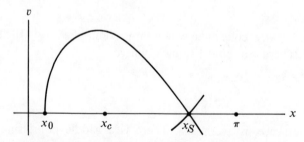

A sketch of the solution curve needed in Stoker's solution of the pull-out problem for a synchronous motor.

We want the solution with a negative slope, so we take

$$a_1 = -\frac{\lambda}{2} - \sqrt{\frac{\lambda^2}{4} - \cos x_S} \;.$$

There is no reason to expect trouble when integrating this solution. All we must do is to properly define $v'(x)$ at x_S in the subroutine for evaluating it.

Stoker's solution of the pull-out problem requires integrating this curve backwards until it vanishes at x_0. The physical quantities desired can all be obtained from x_0. Thus we need to compute a root of the solution, but matters are complicated by the fact that $v'(x_0) = +\infty$. The simplest way to handle this is to interchange the roles of v and x. But we cannot do this until we pass the place where $v'(x) = 0$. One way to proceed is to use STEP to integrate backwards from x_S until, at some mesh point x_n, $v'(x_n) \geq 1$. Then the result v_n at x_n can be used to provide an initial value for

$$\frac{dx}{dv} = \frac{v}{-\lambda v - \sin x + \gamma}, \qquad x(v_n) = x_n,$$

for an integration from v_n to 0. The result is $x(0) = x_0$. This last integration could be done conveniently with DE.

Solve the problem with $\lambda = 0.022$ and $\gamma = 0.8$. You should find that $x_0 \doteq 0.222$.

13

SOLUTIONS TO
THE EXERCISES

The solutions to the exercises are given by chapters. Unless explicitly stated otherwise, all references to equations, theorems, and exercises in the solutions for a chapter are to items in that chapter. For example, in the solutions for Chapter 3, the notation "(6)" refers to equation (6) of Chapter 3. References to an equation (6) in another chapter, such as in Chapter 5, will be made in the form "equation (6) of Chapter 5."

SOLUTIONS FOR CHAPTER 1

1. Because $\cosh(0) = 1$ the function

$$y(x) = \begin{cases} 1 & \alpha \le x \le 0, \\ \cosh(\alpha - x) & x < \alpha, \end{cases}$$

is continuous at $x = \alpha$ hence for all $x \le 0$. Obviously $y(0) = 1$ and

$$y'(x) = \begin{cases} 0 & \alpha \le x \le 0, \\ -\sinh(\alpha - x) & x < a. \end{cases}$$

Since $\sinh(0) = 0$, $y'(x)$ is continuous at $x = \alpha$, hence for all $x \leq 0$. Finally, $y(x)$ satisfies the differential equation:

$$f(x, y(x)) = \begin{cases} -\sqrt{|1 - 1^2|} = 0 & \alpha \leq x \leq 0, \\ -\sqrt{|1 - \cosh^2(\alpha - x)|} = -\sinh(\alpha - x) & x < \alpha, \end{cases}$$

hence $y'(x) = f(x, y(x))$ for all $x \leq 0$.

2. The function

$$y(x) = u(x)[A + v(x)],$$

where

$$u(x) = \exp\left[\int_a^x g(t)\, dt\right], \qquad v(x) = \int_a^x \frac{r(t)}{u(t)}\, dt,$$

is well defined for the following reasons. Since $r(x)$ and $g(x)$ are continuous we note first that $u(x)$ is well defined and positive, and then that $v(x)$ is well defined. Obviously $u(x)$ and $v(x)$ are continuous and have continuous derivatives so $y(x)$ does also.

We first verify that $y(a) = u(a)[A + v(a)] = A$ and then we verify that

$$\begin{aligned} y'(x) &= u(x)v'(x) + u'(x)[A + v(x)] \\ &= r(x) + g(x)u(x)[A + v(x)] \\ &= r(x) + g(x)y. \end{aligned}$$

3. Given that f and f_y are continuous in R and f_y is unbounded there, assume also that f satisfies a Lipschitz condition in R. Then

$$\begin{aligned} \left|\frac{\partial f}{\partial y}\right| &= \left|\lim_{h \to 0} \frac{f(x, y + h) - f(x, y)}{h}\right| \\ &= \lim_{h \to 0} \left|\frac{f(x, y + h) - f(x, y)}{h}\right| \\ &\leq \lim_{h \to 0} \frac{L|h|}{|h|} = L. \end{aligned}$$

But this says f_y is bounded, and the contradiction shows that f cannot satisfy a Lipschitz condition.

4. a) $f_y = \dfrac{1}{1 + x^2}$; any constant $L \geq 1$

   b) $f_y = x$; no Lipschitz constant exists since f_y is unbounded on the region

   c) $f_y = \sin x$; any constant $L \geq 1$

   d) $f_y = \dfrac{1}{2\sqrt{|y|}}$; no Lipschitz constant exists

   e) $f_y = 0$; any constant $L \geq 0$.

5. The choices for $P(x)$, $Q(x)$, $R(x)$, $S(x)$ must satisfy

$$P(x) + S(x) + \frac{Q'(x)}{Q(x)} + g(x) = 0,$$

$$R(x)Q(x) + P'(x) + P^2(x) + P(x)g(x) + h(x) = 0.$$

   a) $Q(x) \equiv 1$, $P(x) \equiv 0$, $R(x) = -h(x)$, and $S(x) = -g(x)$ satisfy these conditions since

$$0 - g(x) + 0 + g(x) = 0,$$

$$-h(x) \cdot 1 + 0 + 0 + 0 + h(x) = 0.$$

   b) $Q(x) = 1/E(x)$, $P(x) \equiv 0$, $S(x) \equiv 0$, $R(x) = -h(x)E(x)$, and $E(x) = \exp\left(\displaystyle\int_a^x g(t)\, dt \right)$ satisfy

$$0 + 0 - \frac{g(x) \cdot E(x)}{E(x)} + g(x) = 0,$$

$$-\frac{h(x)E(x)}{E(x)} + 0 + 0 + 0 + h(x) = 0.$$

   $Q(x)$ is well defined since $E(x)$ is positive for all x.

   c) $Q(x) \equiv 1$, $\quad P(x) = S(x) = -g(x)/2$, $\quad R(x) = g'(x)/2 + g^2(x)/4 - h(x)$ satisfy

$$-\frac{g(x)}{2} - \frac{g(x)}{2} + 0 + g(x) = 0,$$

$$\left(\frac{g'(x)}{2} + \frac{g^2(x)}{4} - h(x) \right) - \frac{g'(x)}{2} + \frac{g^2(x)}{4} - \frac{g^2(x)}{2} + h(x) = 0.$$

6. a) With $y_1 = y$, $y_2 = y'$, the system is

$$y_1' = y_2,$$
$$y_2' = (1 - y_1^2)y_2 - y_1,$$

and requires the initial values $y(a)$, $y'(a)$.

b) With $y_1 = y$, $y_2 = y'$, $y_3 = y''$, the system is

$$y_1' = y_2,$$
$$y_2' = y_3,$$
$$y_3' = \frac{-6x^2y_3 - 7xy_2}{x^3},$$

and requires the initial values $y(a)$, $y'(a)$, $y''(a)$.

c) With $y_1 = x$, $y_2 = \dot{x}$, $y_3 = z$, the system is

$$\dot{y}_1 = y_2,$$
$$\dot{y}_2 = y_2 - \sin y_3,$$
$$\dot{y}_3 = y_1 - y_3$$

and requires the initial values $x(a)$, $\dot{x}(a)$, $z(a)$.

d) With $y_1 = y$, $y_2 = \dot{y}$, $y_3 = v$ we have

$$\dot{y}_1 = y_2,$$
$$\dot{y}_2 + y_2 - \dot{y}_3 = 0,$$

and

$$\dot{y}_2 + \dot{y}_3 + 3y_1 + y_3 = 0.$$

Adding the last two equations we see that

$$2\dot{y}_2 + 3y_1 + y_2 + y_3 = 0;$$

subtracting them we see that

$$2\dot{y}_3 + 3y_1 - y_2 + y_3 = 0.$$

Thus we can solve the system

$$\dot{y}_1 = y_2,$$

$$\dot{y}_2 = -\frac{3y_1 + y_2 + y_3}{2},$$

$$\dot{y}_3 = -\frac{3y_1 - y_2 + y_3}{2}$$

and we require the initial values $y(a)$, $\dot{y}(a)$, $v(a)$.

7. By definition $g_1(h)$ is $O(h^{k_1})$ means there are constants c_1 and h_1 such that

$$|g_1(\dot{h})| \le c_1 h^{k_1} \qquad \text{if } h \le h_1.$$

For any constant α we see that

$$|\alpha g_1(h)| \le |\alpha| c_1 h^{k_1} = ch^{k_1} \qquad \text{if } h \le h_1$$

where we let $c = |\alpha| c_1$. By definition $\alpha g_1(h)$ is $O(h^{k_1})$.

Let c_2 and h_2 be constants such that

$$|g_2(h)| \le c_2 h^{k_2} \qquad \text{if } h \le h_2.$$

Then

$$|g_1(h)g_2(h)| \le c_1 c_2 h^{k_1} h^{k_2} \qquad \text{if } h \le \min(h_1, h_2).$$

If we let $c = c_1 c_2$, $H = \min(h_1, h_2)$, this says that

$$|g_1(h)g_2(h)| \le ch^{k_1 + k_2} \qquad \text{if } h \le H$$

hence $g_1(h)g_2(h)$ is $O(h^{k_1 + k_2})$. In the same way

$$|g_1(h) + g_2(h)| \le c_1 h^{k_1} + c_2 h^{k_2}$$
$$\le \max(c_1, c_2)h^{\min(k_1, k_2)}$$

if $h \le \min(h_1, h_2)$. With $k_3 = \min(k_1, k_2)$ this implies that $g_1(h) + g_2(h)$ is $O(h^{k_3})$.

SOLUTIONS FOR CHAPTER 2

1. All three polynomials simplify to $-2x^2/3 + 11x/3$

2. The augmented difference table is

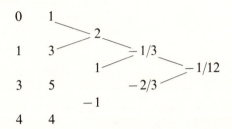

and

$$\bar{P}_4(x) = \bar{P}_3(x) + (x - 4)(x - 3)(x - 1)(-1/12).$$

3. For constant step size h and $m < i$

$$\prod_{\substack{j=0 \\ j \neq m}}^{i} \frac{1}{(x_{k-m} - x_{k-j})} = \frac{-1}{mh} \cdot \frac{-1}{(m-1)h} \cdots \frac{-1}{h} \cdot \frac{1}{h} \cdots \frac{1}{(i-m)h}$$

$$= \frac{(-1)^m}{m!(i-m)!h^i}.$$

The same expression holds for $m = i$. From (5) we see that

$$f[x_k, x_{k-1}, \ldots, x_{k-i}] = \sum_{m=0}^{i} f_{k-m} \prod_{\substack{j=0 \\ j \neq m}}^{i} \frac{1}{x_{k-m} - x_{k-j}}$$

$$= \sum_{m=0}^{i} f_{k-m} \frac{(-1)^m}{m!(i-m)!h^i}.$$

Therefore

$$\nabla^i f_k = i! h^i f[x_k, x_{k-1}, \ldots, x_{k-i}]$$

$$= \sum_{m=0}^{i} (-1)^m f_{k-m} \binom{i}{m}$$

where

$$\binom{i}{m} = \frac{i!}{m!(i-m)!}.$$

SOLUTIONS FOR CHAPTER 3

1. By definition

$$\alpha_{k,i} = \frac{1}{h_{n+1}} \int_{x_n}^{x_{n+1}} l_i(t)\,dt$$

where

$$l_i(x) = \prod_{\substack{j=0 \\ j \neq i}}^{k-1} \left(\frac{x - x_{n-j}}{x_{n-i} - x_{n-j}} \right).$$

With $s = (x - x_n)/h_{n+1}$ this is written

$$\alpha_{k,i} = \int_0^1 l_i(x_n + sh_{n+1})\,ds,$$

$$l_i(x_n + sh_{n+1}) = \prod_{\substack{j=0 \\ j \neq i}}^{k-1} \left(\frac{h_{n+1}s + x_n - x_{n-j}}{x_{n-i} - x_{n-j}} \right).$$

If x is scaled by a factor σ so that x_{n+1-i} is replaced by σx_{n+1-i} for each i, we have h_{n+1} replaced by σh_{n+1}, and

$$\prod_{\substack{j=0 \\ j \neq i}}^{k-1} \left(\frac{\sigma h_{n+1}s + \sigma x_n - \sigma x_{n-j}}{\sigma x_{n-i} - \sigma x_{n-j}} \right)$$

is unaffected. Hence $\alpha_{k,i}$ is unaffected. When the step size is a constant h,

$$l_i(x_n + sh_{n+1}) = \prod_{\substack{j=0 \\ j \neq i}}^{k-1} \left(\frac{hs + jh}{(i-j)h} \right) = \prod_{\substack{j=0 \\ j \neq i}}^{k-1} \left(\frac{s+j}{i-j} \right)$$

is independent of h, as is $\alpha_{k,i}$.

2.

$$\sum_{i=1}^{k} \alpha_{k,i} = \frac{1}{h_{n+1}} \sum_{i=1}^{k} \int_{x_n}^{x_{n+1}} l_i(t)\,dt$$

$$= \frac{1}{h_{n+1}} \int_{x_n}^{x_{n+1}} \sum_{i=1}^{k} l_i(t)\,dt = \frac{1}{h_{n+1}} \int_{x_n}^{x_{n+1}} 1 \cdot dt = 1$$

because $P_{k,n}(t) = \sum_{i=1}^{k} l_i(t) \cdot 1 \equiv 1$, there being only one polynomial of degree $k - 1$ interpolating to $f(x, y) \equiv 1$ on k distinct data points.

The argument for $\sum_{i=0}^{k-1} \alpha_{k,i}^* = 1$ is analogous.

3. From equations (6) and (10) for $i \geq 1$,

$$\gamma_i = \frac{1}{i!} \int_0^1 s(s + 1) \cdots (s + i - 1)\, ds,$$

$$\gamma_i^* = \frac{1}{i!} \int_0^1 (s - 1)(s) \cdots (s + i - 2)\, ds.$$

Then

$$\gamma_i - \gamma_{i-1} = \frac{1}{(i - 1)!} \int_0^1 s(s + 1) \cdots (s + i - 2) \left[\frac{(s + i - 1)}{i} - 1 \right] ds$$

$$= \frac{1}{i!} \int_0^1 (s - 1)(s) \cdots (s + i - 2)\, ds = \gamma_i^*.$$

4. To keep the corrected values straight let us write $y_{n+1}(k)$ for the result of the corrector of order k, and $y_{n+1}(k + 1)$ for that of order $k + 1$. From (5) and (9)

$$y_{n+1}(k) - p_{n+1} = h \sum_{i=1}^{k} \gamma_{i-1}^* \nabla^{i-1} f_{n+1}^p - h \sum_{i=1}^{k} \gamma_{i-1} \nabla^{i-1} f_n$$

$$= h \sum_{i=1}^{k} [\gamma_{i-1}^* - \gamma_{i-1}] \nabla^{i-1} f_{n+1}^p$$

$$+ h \sum_{i=1}^{k} \gamma_{i-1} [\nabla^{i-1} f_{n+1}^p - \nabla^{i-1} f_n]$$

$$= -h \sum_{i=2}^{k} \gamma_{i-2} \nabla^{i-1} f_{n+1}^p + h \sum_{i=1}^{k} \gamma_{i-1} \nabla^i f_{n+1}^p$$

where we use the fact that $\gamma_0 = \gamma_0^* = 1$ and the relation derived in exercise 3. From this last equality it is obvious that

$$y_{n+1}(k) = p_{n+1} + h\gamma_{k-1} \nabla^k f_{n+1}^p.$$

From (9) we see

$$y_{n+1}(k+1) = y_{n+1}(k) + h\gamma_k^* \nabla^k f_{n+1}^P$$
$$= p_{n+1} + h\gamma_{k-1}\nabla^k f_{n+1}^P + h\gamma_k^* \nabla^k f_{n+1}^P$$
$$= p_{n+1} + h\gamma_k \nabla^k f_{n+1}^P.$$

SOLUTIONS FOR CHAPTER 4

1. The order of convergence of the PECE scheme is determined by the order of the truncation error δ, since we assume the starting error E_0 is as small as necessary. For a predictor of order k, the local truncation error τ_{n+1}^P is $O(h^k)$; for a corrector of order m, τ_{n+1} is $O(h^m)$. Since $\delta = \max_n \delta_n = \max_n \left| h\alpha_{m+1,0}^* f_y \tau_{n+1}^P + \tau_{n+1} \right|$ we see that $\delta = O(h^p)$ and $p = \min(m, k+1)$. It follows from Theorem 1 that the PECE scheme is uniformly convergent of order p.

2. By definition

$$y_I'(x) = P_{k+1,n+1}(x),$$

$$y_I(x) = y_{n+1} + \int_{x_{n+1}}^x y_I'(t)\,dt = y_{n+1} + \int_{x_{n+1}}^x P_{k+1,n+1}(t)\,dt$$

where

$$P_{k+1,n+1}(x_{n+2-j}) = f_{n+2-j}, \qquad j = 1,\ldots,k+1.$$

For $k = 1$

$$y_I'(x) = P_{2,n+1}(x) = f_{n+1} + (x - x_{n+1})\left(\frac{f_{n+1} - f_n}{h_{n+1}}\right)$$

and the expression for $y_I(x)$ follows by integrating $y_I'(x)$.

3. The argument is analogous to that leading to Theorem 1. From equation (1)

$$y_{n+1} = y_n + h \sum_{j=1}^k \alpha_{k+1,j}^* f(x_{n+1-j}, y_{n+1-j}) + h\alpha_{k+1,0}^* f(x_{n+1}, p_{n+1})$$

where

$$p_{n+1} = y_n + h \sum_{j=1}^{k} \alpha_{k,j} f(x_{n+1-j}, y_{n+1-j})$$

and there is a similar expression for u_{n+1}. If we define $e_i = y_i - u_i$, then we have

$$|e_{n+1}| \leq |e_n| + hL \sum_{j=1}^{k} |\alpha^*_{k+1,j}| \, |e_{n+1-j}| + hL|\alpha^*_{k+1,0}| \, |e_n|$$

$$+ h^2 L^2 |\alpha^*_{k+1,0}| \sum_{j=1}^{k} |\alpha_{k,j}| \, |e_{n+1-j}|.$$

We simplify this by defining the sequence $\{E_i\}$ such that $|e_j| \leq E_i$ for $j = 0, 1, \ldots, i$ and all i, with

$$E_0 = \max \{|e_0|, |e_1|, \ldots, |e_{k-1}|\}$$
$$= \max \{|y_0 - u_0|, |y_1 - u_1|, \ldots, |y_{k-1} - u_{k-1}|\}.$$

Then

$$E_{n+1} = \mathscr{L} E_n$$

where

$$\mathscr{L} = 1 + hL\alpha^* + h^2 L^2 |\alpha^*_{k+1,0}| \alpha,$$

$$\alpha^* = \sum_{j=0}^{k} |\alpha^*_{k+1,j}|, \qquad \alpha = \sum_{j=1}^{k} |\alpha_{k,j}|.$$

With Lemmas 1 and 2 this implies for any $x_n \in [a, b]$ that

$$|y_n - u_n| \leq E_0 \cdot \exp\left[(b - a)L(\alpha^* + hL|\alpha^*_{k+1,0}|\alpha)\right] = \max_{0 \leq j \leq k-1} |y_j - u_j| \cdot B.$$

4. Assume that the equations $y' = f(x, y)$ satisfy the conservation law $v^T y(x) \equiv c$ for some vector v and constant c. This arises from the identity $v^T f(x, y) \equiv 0$ which holds for *any* y, in particular, for a computed solution y_i. The approximation y_{n+1} generated by a PECE method of order k must satisfy

$$v^T y_{n+1} = v^T y_n + h \sum_{j=1}^{k} \alpha^*_{k+1,j} v^T f(x_{n+1-j}, y_{n+1-j}) + h\alpha^*_{k+1,0} v^T f(x_{n+1}, p_{n+1})$$

where

$$\mathbf{p}_{n+1} = \mathbf{y}_n + h \sum_{j=1}^{k} \alpha_{k,j} \mathbf{f}(x_{n+1-j}, \mathbf{y}_{n+1-j}).$$

Thus with the identity $\mathbf{v}^T \mathbf{f}(x, \mathbf{y}) \equiv 0$, we have $\mathbf{v}^T \mathbf{y}_{n+1} = \mathbf{v}^T \mathbf{y}_n$. The same argument shows $\mathbf{v}^T \mathbf{y}_n = \mathbf{v}^T \mathbf{y}_{n-1}$ and so on, until we have $\mathbf{v}^T \mathbf{y}_{n+1} = \mathbf{v}^T \mathbf{y}_0$. Because we assume exact arithmetic, $\mathbf{y}_0 = \mathbf{y}(x_0)$ hence $\mathbf{v}^T \mathbf{y}_{n+1} = \mathbf{v}^T \mathbf{y}_0 = \mathbf{v}^T \mathbf{y}(x_0) = c$ for all n.

5.

m	$\delta(10^{-m})$	$\Delta(10^{-m})$
1	.3752	3.752
2	.3685	36.85
3	.3679	367.9
4	.3679	3679.

These results demonstrate that a starting error of order h persists and destroys the convergence of order h^2 which the method will yield with an accurate start.

SOLUTIONS FOR CHAPTER 5

1. From the definition (4) it is clear that $\psi_i(n + 1) = ih$ for each i and n if the step size is a constant h. The result of equation (10) in Chapter 2 states that

$$\nabla^{i-1} f_n = (i - 1)! h^{i-1} f[x_n, x_{n-1}, \ldots, x_{n-i+1}]$$

when the step size is constant, hence that

$$\phi_i(n) = \psi_1(n)\psi_2(n) \cdots \psi_{i-1}(n) f[x_n, \ldots, x_{n-i+1}]$$
$$= h \cdot 2h \cdots (i - 1)hf[x_n, \ldots, x_{n-i+1}]$$
$$= \nabla^{i-1} f_n.$$

From the definition of $\beta_i(n + 1)$ we see that

$$\beta_i(n + 1) = \frac{h \cdot 2h \cdots (i - 1)h}{h \cdot 2h \cdots (i - 1)h} = 1 \qquad \text{for } i > 1,$$

hence that

$$\phi_i^*(n) = \beta_i(n + 1)\phi_i(n) = \phi_i(n) = \nabla^{i-1} f_n.$$

For constant step size h, (6) reduces to

$$c_{i,n}(s) = \begin{cases} 1 & i = 1, \\ s & i = 2, \\ \left(\dfrac{sh}{h}\right) \cdot \left(\dfrac{sh + h}{2h}\right) \cdots \left(\dfrac{sh + (i - 2)h}{(i - 1)h}\right) & i \geq 3. \end{cases}$$

Thus for $i \geq 3$

$$\int_0^1 c_{i,n}(s) \, ds = \int_0^1 \left(\frac{s}{1}\right) \cdot \left(\frac{s + 1}{2}\right) \cdots \left(\frac{s + i - 2}{i - 1}\right) ds$$

$$= \gamma_{i-1}$$

as defined in equation (6) of Chapter 3. The cases $i = 1$ and 2 are easily verified.

2. The two expressions are obviously equivalent for $i = 1$ and 2. Assume the expressions are equivalent for $i = m \geq 3$. Then

$$c_{m+1,n}(s) = \left[\alpha_m(n + 1)s + \frac{\psi_{m-1}(n)}{\psi_m(n + 1)}\right] c_{m,n}(s)$$

$$= \left[\frac{h_{n+1}s}{\psi_m(n + 1)} + \frac{\psi_{m-1}(n)}{\psi_m(n + 1)}\right]\left[\left(\frac{sh_{n+1}}{\psi_1(n + 1)}\right) \cdots \left(\frac{sh_{n+1} + \psi_{m-2}(n)}{\psi_{m-1}(n + 1)}\right)\right]$$

and, on rearranging,

$$c_{m+1,n}(s) = \left(\frac{sh_{n+1}}{\psi_1(n + 1)}\right) \cdots \left(\frac{sh_{n+1} + \psi_{m-2}(n)}{\psi_{m-1}(n + 1)}\right) \cdot \left(\frac{sh_{n+1} + \psi_{m-1}(n)}{\psi_m(n + 1)}\right)$$

in agreement with the form (6).

3. By definition

$$c_{1,n}^{(-1)}(s) = \int_0^s 1 \, ds_0 = s$$

and

$$c_{2,n}^{(-1)}(s) = \int_0^s s_0 \, ds_0 = \frac{s^2}{2}.$$

It is obviously the case that the repeated integrals of these expressions are

$$c_{1,n}^{(-q)}(s) = \frac{s^q}{q!}, \qquad c_{2,n}^{(-q)}(s) = \frac{s^{q+1}}{(q+1)!}.$$

Now

$$g_{1,q} = (q-1)! \, c_{1,n}^{(-q)}(1) = (q-1)! \frac{1^q}{q!} = \frac{1}{q},$$

$$g_{2,q} = (q-1)! \, c_{2,n}^{(-q)}(1) = (q-1)! \frac{1^{q+1}}{(q+1)!} = \frac{1}{q(q+1)}.$$

4. From (6) it is obvious that $c_{i,n}(s) > 0$ for $s > 0$ and from this we see that the repeated integrals $c_{i,n}^{(-q)}(1) > 0$ and $g_{i,q} > 0$. Since $\alpha_{i-1}(n+1) > 0$ the relation

$$g_{i,q} = g_{i-1,q} - \alpha_{i-1}(n+1)g_{i-1,q+1}$$

implies $g_{i,q} > g_{i-1,q}$.

5. According to (11)

$$p_{n+1} = y_n + h_{n+1} \sum_{i=1}^k g_{i,1} \phi_i^*(n).$$

From (14) we see that

$$\phi_i^*(n) = \phi_i^p(n+1) - \phi_{i+1}^p(n+1),$$

hence

$$p_{n+1} = y_n + h_{n+1} \sum_{i=1}^{k} g_{i,1} \phi_i^P(n+1) - h_{n+1} \sum_{i=1}^{k} g_{i,1} \phi_{i+1}^P(n+1)$$

$$= y_n + h_{n+1} \sum_{i=1}^{k} (g_{i,1} - g_{i-1,1}) \phi_i^P(n+1) - h_{n+1} g_{k,1} \phi_{k+1}^P(n+1).$$

But, according to (13)

$$y_{n+1} = p_{n+1} + h_{n+1} g_{k+1,1} \phi_{k+1}^P(n+1)$$

$$= y_n + h_{n+1} \sum_{i=1}^{k+1} (g_{i,1} - g_{i-1,1}) \phi_i^P(n+1).$$

When the step size is a constant h we have already noted that

$$\phi_i^P(n+1) = \nabla^{i-1} f_{n+1}^P$$

and $\gamma_{i-1} = g_{i,1}$ hence $g_{i,1} - g_{i-1,1} = \gamma_{i-1} - \gamma_{i-2} = \gamma_{i-1}^*.$ ·

6. By definition

$$y_{\text{out}}' = P_{k+1,n+1}(x_{\text{out}}) = \sum_{i=1}^{k+1} c_{i,n+1}^I(1) \phi_i(n+1).$$

From the recursion

$$c_{i,n+1}^I(1) = c_{i-1,n+1}^I(1) \cdot \Gamma_{i-1}(1)$$

and the definition $\rho_i = c_{i,n+1}^I(1)$, the coefficients ρ_i are given by

$$\rho_1 = c_{1,n+1}^I(1) = 1$$

$$\rho_i = \rho_{i-1} \cdot \Gamma_{i-1}(1) \qquad i = 2, \ldots, k+1.$$

SOLUTION FOR CHAPTER 8

1. The characteristic equation for the Adams-Bashforth method of order k is

$$\zeta^k = \zeta^{k-1} + \lambda \sum_{j=1}^{k} \alpha_{k,j} \zeta^{k-j}$$

and for a PECE method with both formulas of order k,

$$\zeta^k = \zeta^{k-1} + \lambda \sum_{j=1}^{k-1} \alpha_{k,j}^* \zeta^{k-j} + \lambda \alpha_{k,0}^* \left[\zeta^{k-1} + \lambda \sum_{j=1}^{k} \alpha_{k,j} \zeta^{k-j} \right].$$

SOLUTIONS FOR CHAPTER 10

1. A return from DE with $|IFLAG| = 3$ means that the requested accuracy cannot be obtained. Since the code increases the tolerances to values which appear feasible, calling DE again in this way amounts to integrating the problem with the instruction to get the requested accuracy if possible, and otherwise to get all the accuracy that the code believes it can get.

2. Two things are involved in the return on stiffness. First the code must take too many steps (more than 500) and second it must then decide the reason for this is that the equations are stiff. Besides the technical reasons for doing the test in this manner, one simply does not care if the equations are stiff unless they prove to be expensive to integrate. The integration of this example on $[0, 5]$ took 605 function evaluations hence fewer than 303 steps. For this reason no return for stiffness was made in this case.

3. To get a return of $IFLAG = 5$, the integration must take 500 steps and there must be a sequence of at least 50 steps taken at orders no higher than four. When solved as a system one variable is the first derivative of the solution of the original problem. This variable has jumps in its first derivative at multiples of $\pi/2$ so the system has frequent, severe discontinuities. These discontinuities cause the integration to take 500 steps before reaching 20. If DE is used to solve the problem on shorter intervals, for instance $[0, 2]$, $[2, 4]$, ..., $[18, 20]$, no return of $IFLAG = 5$ is seen.

   It should be suspected that a sequence of 50 steps at low orders will be taken near the discontinuity. The discontinuities are very severe with this stringent accuracy request, so the code must go to order one to find and pass them. Using a first order method with a stringent accuracy request makes it likely that a very small step size will be needed at the discontinuity. The solution is very easy to integrate (once past the discontinuity) so the code will double the step size at each step until it reaches a value appropriate to the current order and then work with a constant step size until it can consider raising the order. Thus a long sequence of steps at low orders occurs. Whether or not there will be 50 steps depends on the tolerance and unpredictable matters such as how close the code approaches the discontinuity before going to order one.

4. If $\mathbf{y}' = \mathbf{f}(x, \mathbf{y})$ is integrated from a to b and the code is not forced to stop internally at $x = b$, it will almost surely go beyond b. In this event the integration of $\mathbf{z}' = \mathbf{g}(x, \mathbf{z})$ from b to c cannot be initiated without restarting because the code has already progressed beyond b. If the first integration is stopped internally at b, it is not necessary to set IFLAG $= 1$ for the second integration as the code regards the presence of a new equation at b as a (severe) discontinuity there. The code will locate such a discontinuity and in effect restart itself. The same thing happens if the user programs F(X,Y,YP) to simply start evaluating $\mathbf{g}$ after passing b. It is best for the user to restart the code since it is expensive to locate a discontinuity and the step size and order are not raised as quickly after such a location as it is when the code is told to start. In view of the severity of the discontinuity it is quite possible that the accuracy will be hurt if the user does not restart the code himself.

5. DE attempts to automatically determine the scale of the problem when it selects the initial step size. The initial slopes are used for this purpose but, if the initial derivatives of all solution components vanish (or are very small), the initial step size is taken to be the initial interval. For this problem $y'(0) = \sin 0 = 0$, so the initial step size is π. Ordinarily when the initial step size is far too large, as it is here, the step will fail and the step size will be appropriately adjusted. The code in the first step approximates the derivative of the solution by a constant, and the error estimate is based on the difference between the initial slope and that at the end of the step. Here $y'(\pi) = 0$ so the code believes the derivative has the constant value 0 and that there was no error in the step. This situation arises any time that *all* solution components have derivatives which vanish at the initial point and also vanish at the end of the interval of integration.

NOTATION AND SOME THEOREMS FROM THE CALCULUS

NOTATION

$$\sum_{i=0}^{n} a_i = a_0 + a_1 + \cdots + a_n.$$

$$\prod_{i=0}^{n} a_i = a_0 \times a_1 \times \cdots \times a_n.$$

Computer printout of numbers in scientific notation has two forms. They are exemplified by

$$3.78E - 9 = 3.78D - 9 = 3.78 \times 10^{-9}.$$

The form with "E" comes from single precision numbers, and the form with "D" from double precision numbers.

All vectors $\mathbf{v}$ are printed in boldface. They are column vectors, e.g.,

$$\mathbf{v} = \begin{pmatrix} v_1 \\ v_2 \\ \vdots \\ v_n \end{pmatrix},$$

but it is more convenient to display them as row vectors with the notation of a superscript T to indicate transpose, e.g.,

$$\mathbf{v} = (v_1, v_2, \ldots, v_n)^T.$$

There are a number of equivalent ways of writing both ordinary and partial derivatives:

$$\frac{dy}{dt} = y' = \dot{y}, \quad \frac{d^2y}{dt^2} = y'' = \ddot{y}, \text{ etc.,}$$

$$\frac{\partial}{\partial y} f(x, y) = f_y(x, y), \quad \frac{\partial}{\partial x} f(x, y) = f_x(x, y).$$

$y^{(q)}(s)$, q-th derivative of $y(s)$,

$$y^{(q)}(s) = \frac{d}{ds} y^{(q-1)}(s) = \cdots = \frac{d^q y(s)}{ds^q}.$$

$y^{(-q)}(s)$, q-fold indefinite integral of $y(s)$,

$$y^{(-q)}(s) = \int_0^s y^{(-q+1)}(s_{q-1}) \, ds_{q-1}$$

$$= \int_0^s \int_0^{s_{q-1}} \cdots \int_0^{s_1} y(s_0) \, ds_0 \, ds_1 \ldots ds_{q-1}.$$

Sets are specified by listing all their members, e.g., $\{x_1, x_2, \ldots, x_n\}$, or by specifying a property defining the members. The latter has the form $\{x \mid x \text{ satisfies property } P\}$, and is exemplified by the closed interval

$$[a, b] = \{x \mid a \le x \le b\}$$

and the half-open interval

$$(a, b] = \{x \mid a < x \le b\}.$$

$\in$, member of a set, e.g., $x_2 \in \{x_1, x_2, x_3\}$.
$\rightarrow$, tends to.
$\leftarrow$, replacement operator.
$\doteq$, approximately equal to.
$>$, greater than.
$\gg$, much greater than.

$p(c+)$, the "limit from the right" of $p(x)$ at $x = c$. This notation stands for

$$p(c+) = \lim_{\substack{x \to c \\ x > c}} p(x).$$

There is a similar definition of limit from the left, and the notation for it is $p(c-)$.

$f[x_1, x_2, \ldots, x_k]$, divided difference of order $k - 1$.

$\nabla^k f_n$, backward difference of order k.

$f \in C^k[a, b]$, f belongs to the set of functions with a derivative of order k that is continuous on (a, b) and with a limit from the right at a and a limit from the left at b.

$g(h) = O(h^k)$, g is of order h^k if there are (unknown) constants h_0, C such that $|g(h)| \leq Ch^k$ for all $0 < h \leq h_0$.

THEOREMS

ROLLE'S THEOREM.   Let $f(x)$ be continuous on the finite interval $[a, b]$ and differentiable on (a, b). If $f(a) = f(b) = 0$, there is at least one point ξ such that

$$f'(\xi) = 0 \qquad \text{and} \qquad a < \xi < b.$$

MEAN VALUE THEOREM FOR DERIVATIVES.   Let $f(x)$ be continuous on the finite interval $[a, b]$ and differentiable on (a, b). Then there is at least one point ξ such that

$$\frac{f(b) - f(a)}{b - a} = f'(\xi) \qquad \text{and} \qquad a < \xi < b.$$

MEAN VALUE THEOREM FOR INTEGRALS.   Let $f(x)$ and $g(x)$ belong to $C[a, b]$. If $g(x) \geq 0$ for all x in $[a, b]$, then there is at least one point ξ such that

$$\int_a^b f(x)g(x)\,dx = f(\xi) \int_a^b g(x)\,dx \qquad \text{and} \qquad a < \xi < b.$$

L'HOSPITAL'S RULE.   Let $f(x)$ and $g(x)$ belong to $C^1(a, b)$, and suppose that $f(a+) = 0$ and $g(a+) = 0$. If neither $g(x)$ nor $g'(x)$ vanishes in (a, b), and if

$$\lim_{\substack{x \to a \\ x > a}} \frac{f'(x)}{g'(x)} = A,$$

then

$$\lim_{\substack{x \to a \\ x > a}} \frac{f(x)}{g(x)} = A.$$

TAYLOR'S THEOREM (WITH REMAINDER).   Let $f(x)$ have a continuous derivative of order $n + 1$ on some interval (a, b) containing the points x and x_0. Define $R_{n+1}(x)$ by

$$f(x) = f(x_0) + \frac{f'(x_0)}{1!}(x - x_0) + \frac{f''(x_0)}{2!}(x - x_0)^2 +$$

$$\cdots + \frac{f^{(n)}(x_0)}{n!}(x - x_0)^n + R_{n+1}(x).$$

Then there is at least one point ξ between x and x_0 such that

$$R_{n+1}(x) = \frac{f^{(n+1)}(\xi)}{(n+1)!}(x - x_0)^{n+1}.$$

INTEGRATION BY PARTS.   Let $f(x)$ and $g(x)$ belong to $C^1[a, b]$. Then

$$\int_a^b f(x)g'(x)\,dx = f(b)g(b) - f(a)g(a) - \int_a^b f'(x)g(x)\,dx.$$

The form in which we usually apply this theorem is with b a variable (let us call it s), and

$$g(x) = \int_a^x G(t)\,dt$$

for some continuous function $G(x)$. Then by substitution

$$\int_a^s f(x)G(x)\,dx = f(s)\int_a^s G(x)\,dx - \int_a^s f'(x)\int_a^x G(t)\,dt\,dx.$$

REFERENCES

This text thoroughly treats one numerical method for solving ordinary differential equations and only one implementation of that method; alternate approaches abound in the literature. Surveys of the literature covering the period 1950–1970 are included in the following books.

1. Milne, W. E., *Numerical Solution of Differential Equations,* Dover Publications, New York, 2nd ed., 1970.
2. Henrici, P., *Discrete Variable Methods in Ordinary Differential Equations,* John Wiley & Sons, New York, 1962.
3. Gear, C. W., *Numerical Initial Value Problems in Ordinary Differential Equations,* Prentice-Hall, Englewood Cliffs, N.J., 1971.

The following references are intended to supplement topics discussed in this text.

4. Bailey, P. B., Sturm-Liouville eigenvalues via a phase function, SIAM Journal on Applied Mathematics 14, 1966, 242–249.
5. Bailey, P. B., Shampine, L. F., Waltman, P. E., *Nonlinear Two Point Boundary Value Problems,* Academic Press, New York, 1968.
6. Banks, D. O., Kurowski, G. J., Computation of eigenvalues of singular Sturm-Liouville systems, Mathematics of Computation 22, 1968, 304–310.
7. Birkhoff, G., Rota, G.-C., *Ordinary Differential Equations,* Ginn and Company, New York, 2nd ed. 1969.

8. Conte, S. D., The numerical solution of linear boundary value problems, SIAM Review 8, 1966, 309–321.

9. Franklin, J. N., *Matrix Theory,* Prentice-Hall, Englewood Cliffs, N.J., 1968.

10. Hale, J. K., *Ordinary Differential Equations,* Interscience Publishers, New York, 1969.

11. Hall, G., Stability analysis for the Adams multistep methods of predictor-corrector algorithms of Adams type, SIAM Journal on Numerical Analysis 11, 1974, 494–505.

12. Henrici, P., *Error Propagation for Difference Methods,* John Wiley & Sons, New York, 1963.

13. Hull, T. E., Enright, W. H., Fellen, B. M., Sedgwick, A. E., Comparing numerical methods for ordinary differential equations, SIAM Journal on Numerical Analysis 9, 1972, 603–637.

14. Keller, H., *Numerical Methods for Two-Point Boundary-Value Problems,* Blaisdell, Waltham, Mass., 1968.

15. Klamkin, M. S., Transformation of boundary value problems into initial value problems, Journal of Mathematical Analysis and Applications 32, 1970, 308–330.

16. Krogh, F. T., A test for instability in the numerical solution of ordinary differential equations, Journal of the Association for Computing Machinery 14, 1967, 351–354.

17. ———, Algorithms for changing the step size, SIAM Journal on Numerical Analysis 10, 1973, 949–965.

18. ———, Changing step size in the integration of differential equations using modified divided differences, Proceedings of the Conference on the Numerical Solution of Ordinary Differential Equations, Lecture Notes in Mathematics No. 362, Springer-Verlag, New York, 1974.

19. ———, On testing a subroutine for the numerical integration of ordinary differential equations, Journal of the Association for Computing Machinery 20, 1973, 545–562.

20. ———, VODQ/SVDQ/DVDQ—variable order integrators for the numerical solution of ordinary differential equations, TU Doc. No. CP-2308, NPO-11643, May 1969, Jet Propulsion Laboratory, Pasadena, Ca.

21. Lefschetz, S., *Differential Equations: Geometric Theory,* Interscience Publishers, New York, 1957.

22. Osborne, M. R., On shooting methods for boundary value problems, Journal of Mathematical Analysis and Applications 27, 1969, 417–433.

23. Roberts, S. M., Shipman, J. S., *Two-Point Boundary Value Problems: Shooting Methods,* Elsevier, New York, 1972.

24. Sánchez, D. A., *Ordinary Differential Equations and Stability Theory: an Introduction,* W. H. Freeman and Company, San Francisco, 1968.

25 Shampine, L. F., Allen, R. C., *Numerical Computing: an Introduction,* Saunders, Philadelphia, 1973.

26. Stetter, H. J., *Analysis of Discretization Methods for Ordinary Differential Equations,* Springer-Verlag, New York, 1973.

27. Stoker, J. J., *Nonlinear Vibrations,* Interscience Publishers, New York, 1950.

28. Wilkinson, J. H., *Rounding Errors in Algebraic Processes,* Prentice-Hall, Englewood Cliffs, N.J., 1963.

INDEX